General Submissions

AA008946

Held at the 2023 AIChE Annual Meeting

Orlando, Florida, USA
5-10 November 2023

ISBN: 978-1-7138-9302-8

Printed from e-media with permission by:

Curran Associates, Inc.
57 Morehouse Lane
Red Hook, NY 12571

Some format issues inherent in the e-media version may also appear in this print version.

Printed with permission by Curran Associates, Inc. (2024)

For permission requests, please contact AIChE
at the address below.

AIChE
120 Wall Street, FL 23
New York, NY 10005-4020

Phone: (800) 242-4363
Fax: (203) 775-5177

www.aiche.org

Additional copies of this publication are available from:

Curran Associates, Inc.
57 Morehouse Lane
Red Hook, NY 12571 USA
Phone: 845-758-0400
Fax: 845-758-2633
Email: curran@proceedings.com
Web: www.proceedings.com

TABLE OF CONTENTS

SYNTHETIC NEUROBIOLOGY

SYNTHETIC CELLS AND CELL FREE SYSTEMS

BIOMOLECULES: COMPUTATION, DESIGN AND EVOLUTION

MULTICELLULAR AND ORGAN-SCALE ENGINEERING

BIOLOGICAL INFORMATION STORAGE AND COMPUTING

THE FUTURE OF FOOD AND NATURAL MEDICINES

BIG-DATA AND SYNTHETIC BIOLOGY

ENGINEERED LIVING MATERIALS

POSTER PARTICIPANTS

Genetically Programmable Barcodes for Correlative Volume Electron Microscopy

Felix Sigmund, Oleksandr Berezin, Sofia Beliakova, Bernhard Magerl, Martin Drawitsch, Alberto Piovesan, Filipa Gonçalves, Silviu-Vasile Bodea, Stefanie Winkler, Zoe Bousraou, Martin Grosshauser, Eleni Samara, Jesús Pujol-Martí, Sebastian Schädler, Chun So, Stephan Irsen, Axel Walch, Florian Kofler, Marie Piraud, Joergen Kornfeld, Kevin Briggman, Gil Gregor Westmeyer

Multi-color fluorescent proteins have revolutionized biomedical research delivering real-time insights into dynamic cellular processes in living cells. For Electron Microscopy (EM) with its unparalleled resolution, to date, there are still few multiplexable genetic tools and existing ones usually require the addition of exogenous substrates during extended workflows.

To overcome this limitation, we have previously introduced iron-mineralizing Encapsulins, a class of prokaryotic capsid-like nanocompartments, as genetic tools for multi-modal molecular imaging including cryo-electron tomography (cryo-ET) and standard EM[1, 2].

In this current work[3], we engineered an extended array of spherically symmetric, concentric barcodes ("EMcapsulins") compatible with standard EM workflows (TEM and (FIB)-SEM) by fusing concatenated metal-interacting protein domains of varying lengths to Encapsulin scaffolds of different diameters. Six different classes could be automatically detected and classified using common deep-learning architectures. The coding capacity could be further augmented by combining differently sized EMcapsulins with rigid, elongated linkers of varying lengths yielding highly multiplexable nano-patterns. Furthermore, EMcapsulins could be rendered fluorescent for correlative workflows and a framework for tunable stoichiometry control of surface binders was implemented enabling non-invasive labeling of subcellular protein targets. Neuronal expression of EMcapsulins in *Drosophila* and mouse brain enabled the automated, multiplexed identification of a genetically defined subset of cells. EMcapsulins can augment (sub)nanometer resolution anatomical EM data with multiplexed molecular information. Engineering of EMcapsulins responsive to the levels of cellular analytes is currently ongoing and will enable their spatial mapping at EM resolution.

1. Sigmund, F. *et al.* A. Bacterial encapsulins as orthogonal compartments for mammalian cell engineering. *Nat. Commun.* **9**, 1990 (2018).

2. Farhadi, A., Sigmund, F., Westmeyer, G. G. & Shapiro, M. G. Genetically encodable materials for non-invasive biological imaging. *Nat. Mater.* **20**, 585–592 (2021).

3. Sigmund, F. *et al.* Genetically encoded barcodes for correlative volume electron microscopy. *Nat. Biotechnol.* (2023) (in press, https://doi.org/10.1038/s41587-023-01713-y)

Automated Tranformation Enabling Automated Genome Engineering

Peter Ruppen, Michael Gerlt, Sven Panke

Genetic engineering allows us to fundamentally change organisms such as bacteria and develop novel solutions to the world's problems. A key process in genetic engineering protocols is the delivery of DNA into the cells (transformation). However, this step involves multiple laborious and time-consuming manual steps. To address this issue, we have developed a microfluidic system that simplifies and accelerates bacterial transformation. Our automated system enables transformation with the click of a button.

The system consists of two interconnected devices: an acoustic medium exchanger and a flow-through electroporator, mimicking the two steps performed in manual processing. In the acoustic medium exchanger acoustic streaming and acoustic radiation forces are used to first capture bacteria *(Escherichia coli)* suspended in growth medium. Subsequently, the flow is switched to deionized water to wash away the growth medium. Finally, DNA is added, and the acoustics are turned off and the cells leave the medium exchanger and directly flow through the second device, the electroporator, where transformation is performed.

The electroporation device consists of two titanium capillary electrodes that are separated by a thin PTFE disk with a small hole that aligns with the capillaries' holes. The arrangement of the electrodes leads to a uniform electric field inside the disk. The field is proportional to the applied voltage and inversely proportional to the distance between the electrodes, which is equivalent to the thickness of the PTFE disk.

This microfluidic design allows for the transformation in small volumes, which is traditionally difficult due to problems with centrifugation of small number of cells. The benefit of utilizing small volumes is particularly advantageous when performing evolution on substrates that are difficult to synthetize in large amounts.

Developing Artificial Cell Micro-Factories to Support Tissue Regeneration

Gal Chen

Recent advances in bottom-up synthetic biology have catalyzed the development of synthetic cells (SCs), autonomous protein-manufacturing particles, as potential biomimetics for replacing diseased natural cells and addressing medical needs. By leveraging the unique properties of cell nature, synthetic cell technologies aim to create well-defined, regulated, and advanced drug delivery systems. In this study, we demonstrate that SCs, genetically engineered to produce proangiogenic factors, induced the physiological process of neovascularization in mice. The SCs were created from giant lipid vesicles and were optimized for enhanced protein production. Upon introducing the appropriate genetic code, the SCs synthesized a recombinant human basic fibroblast growth factor (bFGF), with expression levels reaching up to $9 \cdot 10^6$ protein copies per SC. In cell culture, the SCs stimulated endothelial cell proliferation, migration, tube formation, and intracellular signaling related to angiogenesis, confirming their proangiogenic capabilities. When integrated with bioengineered constructs containing endothelial and mural cells, the SCs promoted the remodeling of stabilized vascular networks, supported by a collagen-IV basement-membrane-like matrix. In vivo, sustained local administration of the SCs in mice activated the infiltration of blood vessels into implanted Matrigel plugs with no detected systemic immunogenicity. These results underscore the potential of SCs as therapeutic platforms that autonomously produce biological drugs inside the body, stimulating physiological processes and treating medical conditions.

Experimental evolution of chosen genes near their mutational speed limits

Chang Liu

My laboratory engineers genetic systems that go beyond what nature's genetic systems can do. We are especially interested in creating genetic systems that dramatically accelerate the speed of evolution and that record transient information as interpretable mutations in DNA. We apply these systems to the generation of useful biomolecules and therapeutics; the study of molecular evolution; and cell and developmental biology. I will discuss recent work on our orthogonal DNA replication system for the continuous evolution of biomolecules in vivo, including new DNA polymerases that further accelerate the speed of chosen gene evolution and their applications to enzyme and antibody engineering. I will also offer ideas for uncovering 'secrets' of evolution including broad sampling of fitness landscapes through the generation of diverse evolutionary datasets and prospectively testing how evolutionary pasts influence futures.

Phage and Robotics-Assisted Directed Evolution of Engineered Biomolecules

Emma Chory

Evolution occurs when selective pressures from the environment shape inherited variation over time. Within the laboratory, evolution is commonly used to engineer proteins and RNA, but experimental constraints have limited our ability to reproducibly and reliably explore key factors such as population diversity, the timing of environmental changes, and chance. We developed a high-throughput system for the analytical exploration of molecular evolution using phage-based mutagenesis to evolve many distinct classes of biomolecules simultaneously. In this talk, I will describe the development of our open-source python:robot integration platform which enables us to adjust the stringency of selection in response to real-time evolving activity measurements and to dissect the historical, environmental, and random factors governing biomolecular evolution. Finally, I will talk about our many on-going and future projects which utilize this system to evolve previously intractable biomolecules to target the undruggable proteome of protein-protein interactions.

A Synthetic Protein-Level Neural Network in Mammalian Cells

Zibo Chen, JM Linton, Ronghui Zhu, Michael Elowitz

Artificial neural networks provide a powerful paradigm for information processing that has transformed diverse fields. Within living cells, genetically encoded synthetic molecular networks could, in principle, harness principles of neural computation to classify molecular signals. Here, we combine de novo designed protein heterodimers and engineered viral proteases to implement a synthetic protein circuit that performs winner-take-all neural network computation. This "perceptein" circuit includes modules that compute weighted sums of input protein concentrations through reversible binding interactions, and allow for self-activation and mutual inhibition of protein components using irreversible proteolytic cleavage reactions. Altogether, these interactions comprise a network of 310 chemical reactions stemming from 8 expressed protein species. The complete system achieves signal classification with tunable decision boundaries in mammalian cells. These results demonstrate how engineered protein-based networks can enable programmable signal classification in living cells.

Self-Driving Laboratories to Autonomously Navigate the Protein Fitness Landscape

Jacob Rapp, Bennett Bremer, Philip Romero

Human researchers engineer biological systems through the iterative process of generating hypotheses, designing experiments, performing the experiments in the wet laboratory, and interpreting the resulting data to converge on knowledge of biological mechanism and design new systems with enhanced properties and behaviors. We sought to build a fully autonomous system to mimic and accelerate the human biological discovery and design process. We introduce the *Self-driving Autonomous Machines for Protein Landscape Exploration* (SAMPLE) platform to rapidly engineer proteins without human intervention, feedback, or bias. SAMPLE is driven by an intelligent agent that learns protein sequence-function relationships from data and designs new proteins to test hypotheses. The agent interacts with the physical world though a fully automated robotic system that experimentally tests designed proteins by synthesizing genes, expressing proteins, and performing biochemical measurements of enzyme activity.

We applied SAMPLE with the goal of navigating and optimizing the glycoside hydrolase (GH1) thermostability landscape. We deployed four independent SAMPLE agents that each explored diverse search trajectories, but all converged to similar regions of the protein fitness landscape. Each agent discovered GH1 sequences that were at least 12 °C more stable than the initial natural starting sequences. SAMPLE agents actively acquire information in a new environment, and once they gain confidence in the underlying landscape, will begin to design highly fit protein sequences. SAMPLE is a general-purpose platform that can be readily and broadly applied to diverse classes of proteins and enzymes. Self-driving laboratories will revolutionize the fields of biomolecular engineering and synthetic biology by automating and accelerating highly inefficient, time consuming, repetitive, and laborious biological engineering experiments.

Evolution of Efficient Biocatalysts That Use Artificial Redox Cofactors

Han Li

Biological production of chemicals often requires the use of redox cofactors, such as nicotinamide adenine dinucleotide (phosphate) (NAD(P)). These cofactors are expensive to use *in vitro* and difficult to control *in vivo*. Simpler, artificial redox cofactors are much cheaper alternatives to NAD(P) *in vitro*, and can deliver reducing power in an orthogonal manner *in vivo*. However, the bottleneck is to engineer enzymes that can efficiently use these artificial cofactors. We developed two universal, growth-based selection platforms to evolve enzymes that can oxidize or reduce nicotinamide mononucleotide (NMN), a model artificial redox cofactor, respectively. This is achieved by metabolically engineering two auxotroph *E. coli* strains: one whose glycolysis requires the oxidized NMN to function, and the other requires the reduced NMN (NMNH) to combat oxidative stress. The evolved enzymes, a phosphite dehydrogenase (PTDH) and a water-forming oxidase (Nox), supported an industrially viable total turnover number (TTN) of ~45,000 with NMN in cell-free biotransformation, the highest TTN for artificial cofactor reported to date. Next, by wiring PTDH into Ecoli's life-essential phosphate assimilation, we evolve this enzyme to recycle a range of other, even simpler artificial cofactors such as 1-benzylnicotinamide (BNA) and 1-Butylnicotinamide (BuNA). Surprisingly, the mutations selected do not directly interact with the artificial cofactors. Instead, biochemical and computational studies revealed a general design principle where the mutations mimic the larger, natural cofactors NAD(P), in modulating the enzymes' conformational dynamics. In summary, although natural evolution has only explored NAD and NADP as two universal redox cofactors in all life, we demonstrate that an expanded toolkit of artificial redox cofactors can be developed in laboratory evolution for efficient biocatalysis.

Phloretin: A Novel Synthetic Mammalian Cell-Cell Communication System

Katherine Kiwimagi, Jean Disset, Timothy Jones, Xavier Duportet, Ashiwarya Jagtap, Christopher Doering, Joseph Berleant, Carlos Barajas, Domitilla Del Vecchio, Ron Weiss

Here we present the development of a novel sender to receiver cell system based on creation and destruction of the small molecule, Phloretin. To find a base set of parts for construction of the system, we searched the literature across multiple species including several plants and bacteria, to find components capable of interacting with small molecules. Considering the large range of interactions already possible with Phloretin we selected a base set of molecules and broke them down into groups for sender, receiver and degrader cell function. For each functional set we performed an optimization step via synthetic design, mutation and/or component balancing. We characterized our system to show fold change on the order of 10 between sender and receiver components and degradation of greater than 95% percent of signal. To show spatiotemporal control we created a self-generated source-based gradient in mammalian cells with a characteristic length measurable at 73 hrs and stable through 85 hrs. We also created the first synthetic sink-based gradient in mammalian cells with a characteristic length measurable at 72 hrs and stable through 84hrs. To help elucidate the governing biophysics for both our sink and source-based gradients we developed a mathematical model capable of recapitulating system behavior as measured by a defined set of metrics. Our model revealed a base set of reactions governing aspects of programalibly in these gradients, including phloretin sequestration by cell membranes. Our work presents a new programmable hormone-like cell to cell communication system for mammalian cells.

Genetic Circuits That Differentiate Microbial Cells into Communities with Distributed Functions

Arash Farhadi, Juliana R. Abramson, Christopher A. Vaiana, Christopher A. Voigt

Single cell organisms can perform many interesting functions as a community by distributing their collective tasks among different members of their population. This division of labor allows cells to collectively perform many tasks without burdening each individual cell with all the functions. There is great interest in developing robust and efficient genetic circuits that can similarly distribute tasks across cells in a population as we engineer microbes to perform many functions for our benefit. To achieve this, we have developed a genetic circuit, inspired by the hierarchical lineage differentiation of stem cells, which allows engineered cells to differentiate from a monoclonal progenitor cell type into a multi-member community with distributed tasks across the population. The genetic circuit uses the DNA rearrangement reactions of serine integrases to differentiate the state of cells. Initially, cells are engineered to be in the progenitor state, with all the promoters that drive the output functions turned off. Upon activation of the differentiation circuit, the encoded integrase-logic stochastically differentiates the circuit to activate only one promoter, allowing the cell to produce a single output function. Cells engineered with these circuits can be maintained for many generations in the quiescent progenitor state, and once activated, in the differentiated community. The topology of the differentiation landscape can be finely tuned to achieve a desired relative ratio of states from 5-90%. In addition, we have developed up to four member communities, and show a roadmap to scaling up the number of possible states using a combination of orthogonal integrase attachment sequences and ECF sigma factors. The division of labor genetic circuits will enable engineered cells to more effectively produce multiple resource-intensive products in challenging environments such as the human and plant-associated microbiomes.

Exploiting unique molecular through system level properties of DNA as information technologies

Albert Keung

DNA holds significant promise as a data storage medium and computational substrate due to its density, longevity, and resource and energy sustainability. These advantages arise from the inherent biomolecular structure of DNA as well as, perhaps unintuitively, the spatially disordered nature of DNA mixtures. For example, disordered pools of DNA strands confer extreme density, while arraying DNA strands on fixed substrates or within microwells would abrogate almost all of this density advantage. Furthermore, even just terabyte-sized databases would require mixtures of billions of distinct DNA strands, much larger than the sequence diversity of the human genome; thus, the unique molecular and disordered nature of DNA storage presents clear challenges arising simply from the fact that many diverse strands of DNA will be in crowded proximity with severe thermodynamically-driven limitations on file access specificity and accuracy. This prompts important discussions on how data should be organized, accessed, and manipulated. Here we argue that the novel architecture of DNA storage may actually confer novel functionalities beyond what is possible with conventional storage media, and that inspiration from natural biological processes and biochemistry can be leveraged to engineer novel and scalable functionalities into DNA databases. We will describe three examples of this approach. We will show how hierarchical file systems can be engineered to expand the number of possible unique DNA file addresses by five orders of magnitude and that individual KB sized files can be specifically accessed from large TB sized background databases. Next, we took inspiration from the processes of gene expression and strand displacements to engineer a dynamic and reusable DNA storage system that can rename, lock, and delete files while also improving encoding efficiency and information density over current random access methods. Finally, we describe how thermodynamics can be leveraged to tune biomolecular interactions and implement functions such as File Preview.

Distributed Computing of a 2-Bit Hashing Algorithm across an *E.coli* Population

Jai Padmakumar, Jessica Sun, Christopher A. Voigt

Single cells can be programmed to execute a desired function using genetic circuits. However, the size of the genetic circuit that can function in one cell is constrained by the cell's limited resources. To scale genetic circuit designs, we can engineer a population of cells to perform powerful distributed computing without over-burdening one cell. Here, we introduce a method to systematically partition a large circuit across cells as well as the use of phage repressor logic gates and orthogonal cell-cell chemical communication channels to build each genome-integrated subcircuit. We demonstrate the functionality of the method by constructing a 16-input, 2-output genetic circuit across multiple cells, where the genetic circuit performs a single iteration of an 8-bit hash algorithm. The implemented algorithm is related to MD5, a widely used cryptographic function and predecessor to cryptographic functions underlying cryptocurrency. The algorithm implementation uses 110 logic gates partitioned over 65 strains of *Escherichia coli*, requiring 0.62Mb of synthetic DNA to be integrated into their genomes. Each strain was experimentally verified to confirm that they processed their assigned input signals correctly and could propagate the result to the downstream cell. This work marks an improvement in the scale of genetic circuit designs, which brings us closer to harnessing the computational power of cells to implement highly parallelized algorithms and solve computationally difficult tasks.

DNA Cryptography in Living Cells through Massively Parallel Base Editing

Verena Volf, Simon Zhang, George M. Church, Fei Chen

DNA is increasingly used as a medium for information transfer and storage, but methods for security and authentication remain to be addressed. Here, we demonstrate a cryptographic encoding scheme, Genomic Sequence Encryption (GSE), in the DNA of living cells which is based solely on the properties of DNA sequencing and orthogonal to existing algorithmic encryption approaches. We implement message encoding through massively parallel Cas9-base editing and demonstrate reliable retrieval through targeted sequencing. Next, we demonstrate, through experiments and computational simulations, the asymmetrical difficulty of detecting encrypted information for an adversary who is not able to perform targeted sequencing, analogous to asymmetric costs in computational cryptography. Importantly, we also show that a minimum number of information transfers need to be observed before an adversary can break the encryption even if unlimited cost is assumed. Lastly, we conceptualize and demonstrate an authentication scheme to ensure the verification of messages, and authentication of living organisms using GSE. Thus, this work represents a completely biological instantiation of many key concepts in modern cryptography in biological substrates.

Engineering Next-Generation T Cells for Cancer Immunotherapy

Yvonne Chen

The adoptive transfer of T cells expressing chimeric antigen receptors (CARs) has demonstrated clinical efficacy in the treatment of advanced cancers, with anti-CD19 CAR-T cells achieving up to 90% complete remission among patients with relapsed B-cell malignancies. However, challenges such as antigen escape and immunosuppression limit the long-term efficacy of adoptive T-cell therapy. Here, I will discuss the development of and clinical data on next-generation T cells that can target multiple cancer antigens and resist antigen escape. I will also present recent work on tuning CAR signaling activities via rational protein design to achieve greater *in vivo* anti-tumor efficacy. This presentation will highlight the potential of synthetic biology in generating novel mammalian cell systems with multifunctional outputs for therapeutic applications.

Structurally-Informed Design of T Cell Receptor Specificity Using Deep Learning to Target Peptide-MHC Complexes

Amir Motmaen, David Feldman, Anastasia Minervina, William White, Basile Wicky, Lukas Milles, Mohamad Abedi, Jeremiah Sims, Paul Thomas, David Baker, Phil Bradley

The ability to rewire T cell responses to peptide-MHC complexes of interest will revolutionize the development of the next generation of personalized and off-the-shelf immunotherapies for viral diseases and cancer. To redesign native T cell receptors (TCRs) to bind peptide-MHC antigens of interest, we developed a structure-based interface design method based on iterative structure prediction using AlphaFold and ProteinMPNN sequence design. Our method leverages natural TCR sequence diversity to explore the structural space of their complementary determining regions (CDR) at the peptide-MHC complex interface to rationally redesign the CDRs. To validate our re-designed TCRs, we developed an arrayed in vitro screening method to assess binding and functionality of our TCRs in Jurkat cells. Utilizing our design and screening pipeline, we have already identified re-designed TCRs that efficiently bind the LCMV-gp33 epitope despite the highly divergent amino acid sequences of our designed CDRs compared to the native TCR. Moreover, we have extended our in vitro TCR validation method to enable pooled assembly, packaging, and NGS-based screening of both TCR binding and activity against peptide-MHCs of interest. We believe that our rational design and high-throughput screening method not only enable extensive engineering of TCR specificity for developing therapeutics but also will facilitate in depth characterization of TCR function in sequence and structure space to answer long-standing questions regarding T cell recognition.

Towards DNA-Free Synthetic Biology: Protein Circuits and RNA Sensors

Xiaojing Gao

Synthetic biomolecular circuits hold the promise of transforming therapeutics with sense-and-response capabilities, while mRNA has emerged as a promising delivery vector. However, historically most synthetic circuits rely on DNA-level regulation and are therefore incompatible with RNA delivery. I will share our recent progress, both published and preliminary, in platforms that enable protein- and RNA-level regulation. We established protease-controlled secretion and cell-surface display of cargo proteins, and we demonstrated its performance in mice. We continue to improve the platform using a massively parallel assay, and we have also developed an even more compact design. In parallel, we leveraged RNA-editing, mediated by an endogenous enzyme, to achieve programmable RNA sensing in living cells. The sensor already shows the potential to function robustly in diverse cellular contexts and against diverse RNA inputs, and our next-generation design has further expanded the latter aspect.

Synthetic Genetic Circuits to Uncover and Enforce the OCT4 Trajectories of Successful Reprogramming of Human Fibroblasts

Domitilla Del Vecchio

Reprogramming human fibroblasts to induced pluripotent stem cells (iPSCs) is inefficient, with heterogeneity among transcription factor (TF) trajectories driving divergent cell states. Nevertheless, the impact of TF dynamics on reprogramming efficiency remains uncharted. Here, we identify the successful reprogramming trajectories of the core pluripotency TF, OCT4, and design a genetic controller that enforces such trajectories with high precision. By combining a genetic circuit that generates a wide range of OCT4 trajectories with live-cell imaging, we track OCT4 trajectories with clonal resolution and find that a distinct constant OCT4 trajectory is required for colony formation. We then develop a synthetic genetic circuit that yields a tight OCT4 distribution around the identified trajectory and outperforms in terms of reprogramming efficiency other circuits that less accurately regulate OCT4. Our synthetic biology approach is generalizable for identifying and enforcing TF dynamics for cell fate programming applications.

A Cellular Programming Platform That Integrates Decision-Making, Memory, and Communication *Via* Synthetic Genetic Circuits

Brian Huang, Dowan Kim, Corey Wilson

The development of synthetic, intelligent biological systems has largely focused on three fundamental aspects of cellular behavior— decision-making, memory, and communication. Here, we report a cellular programming platform that leverages modular recombinase-based genetic circuits to facilitate the seamless integration of these three attributes of cellular intelligence. Genetic programs are executed in engineered *Escherichia coli* strains harboring a genome-integrated array of six inducible recombinases. Programs are composed from well-defined memory blocks in the form of gain-of-function and loss-of-function recombination circuits. We then introduce a means to both expand the capacity of memory circuits and couple them with advanced decision-making *via* the development of CRISPR Cas9-mediated protection (CRISPRp) of recombinase action. Next, we demonstrate how intercellular communication can be incorporated into programs through autoinduction and synthetic quorum sensing circuits. Finally, to demonstrate the utility of this platform technology, we show how a blood biosensor can be used to trigger information exchange between probiotic *E. coli* and the commensal gastrointestinal model organism *Bacteroides thetaiotaomicron*.

Development of Synthetic Genetic Circuits to Improve Water-Use Efficiency in Sorghum

Janina Tamborski, Jennifer A.N. Brophy

Sorghum bicolor is a biofuel crop that offers great potential because of its tolerance to drought, heat and low-cost of production as compared to other potential feedstocks. Despite yield gains through breeding, its productivity in suboptimal conditions is still limited. To avoid competition with food crops for water and land, we aim to improve Sorghum water use efficiency through genetic engineering. Water use efficiency is a complex trait, influenced by plant root architecture, leaf cell patterning, and photosynthetic efficiency and synthetic genetic circuits allow us to precisely modify each of these plant features. Although there has been significant progress in developing tools and modifying model plants, this knowledge does not always transfer to crop plants. In the past, we successfully used synthetic genetic circuits in combination with tissue-specific promoters to modify root architecture in *Arabidopsis thaliana*. We are now testing and optimizing these tools for the application in grass species. Through the use of transient protoplast expression in the model bioenergy grass *Setaria viridis*, we show that our parts can be modified in multiple aspects to drive gene expression in a predictable manner. Subsequently, we successfully built genetic circuits that implement NOT IMPLIES logic in grasses. Our initial results indicate a promising future for engineering complex traits in bioenergy grasses.

Towards Universal Synthetic Heterotrophy in Baker's Yeast.

Vikas Trivedi, Sean F. Sullivan, Venkatesh Endalur Gopinarayanan, Taylor Hart, Debika Choudhury, Nikhil Nair, Nikhil U. Nair

The exploration of non-model yeasts for biomanufacturing applications is partly driven by the difficulty in engineering *Saccharomyces cerevisiae* for utilization of non-native but abundant substrates. Despite decades of effort, there has been no systematic analysis to identify the root cause behind the recalcitrance of this yeast for synthetic heterotrophy and how to overcome it for any non-native substrate. Through a combination of directed evolution, flux balancing, synthetic biology, systems biology, and network analysis, we find that the major limitation to substrate utilization is, extrinsic – genes that control the flux of the substrate into central carbon metabolism – and not any downstream (intrinsic) yeast-specific factor. We show that this holds true for structurally diverse substrates, which is a remarkable finding since the current assumption amongst researchers is that the limitations are largely intrinsic, and that each substrate needs a unique engineering strategy. Thus, **our work turns the current long-standing paradigm on its head**. We also show that by perturbing intrinsic factors, cells become less robust and are more sensitive to growth inhibitors commonly found in biomass hydrolysates. We conclude that much of the engineering approach insofar in the literature results in unnecessary and that strains end up being over-engineered with poor growth rates and/or robustness. Our described minimalistic, holistic engineering approach is more rapid, easier to optimize, and results in strains that maximize growth and robustness phenotypes.

Associated works: Endalur Gopinarayanan & Nair (2018) Nat Commun, Trivedi et al. (2022) bioRxiv, Sullivan et al. (2022) bioRxiv.

A Genetically-Engineered Probiotic Designed to Consume Methionine for the Treatment of Homocystinuria

Analise Reeves

Homocystinuria (HCU) is an autosomal recessive disorder characterized by an accumulation of homocysteine (Hcy) in the plasma caused by loss of function of cystathionine beta-synthase. Maintaining a tightly controlled methionine-restricted diet is the first-line, and often only treatment available to HCU patients to avoid multisystem clinical manifestations, including thromboembolism, ectopia lentis and intellectual disability. Synthetic Biotics combine a probiotic *E. coli* Nissle 1917 (EcN) chassis with precision genetic engineering to target and degrade harmful metabolites in the gastrointestinal tract of patients. SYNB1353 is a Synthetic Biotic designed to metabolize methionine in the gut as a mechanism to lower plasma homocysteine levels in patients with HCU. The strain expresses an importer to internalize methionine and an optimized methionine decarboxylase (MetDC) to degrade methionine into 3-methylthiopropylamine (3-MTP). The initial prototype design was optimized by screening both metagenomic and protein engineered libraries of approximately 1000 decarboxylases and nearly 160 importers. The top performing, optimized enzymes and transporters were ultimately combined and integrated into SYNB1353, which degrades methionine and produces 3-MTP at a significantly greater rate than wild type EcN. In a preclinical, acute non-human primate model of HCU, SYNB1353 lowered methionine and Hcy levels in the plasma, confirming the strain is active and metabolizes methionine in vivo. A phase 1 clinical study conducted in healthy volunteers given an oral methionine challenge demonstrated that SYNB1353 was well tolerated and reduced plasma methionine by 26%. Overall, SYNB1353 represents a promising approach for methionine reduction with potential utility for the treatment of HCU and represents one of the first examples of in-human studies with a live biotherapeutic as a new modality for the treatment of disease.

Bacterial Deodorant: Engineering Human Skin Commensals to Degrade Volatile Mosquito Attractants from the Skin Metabolome

Daniel Yoo, Christopher A. Voigt

Mosquitoes are a major source of vector-borne diseases, including malaria, yellow fever, Zika, dengue, and West Nile virus, affecting millions globally each year. Host-seeking mosquitoes are primarily guided by chemical cues released from their blood hosts with individuals displaying significant differences in relative attractiveness to mosquitoes due to various factors, i.e. skin bacterial abundance and diversity. Analysis of human skin emanations often result in hundreds of compounds as potential attractants. By employing synthetic biology tools for primary isolates from the human skin microbiome, we have engineered commensal bacterial strains to degrade or convert volatile mosquito attractants through the successful genomic integration of custom enzymatic pathways. These pathways are capable of degrading 85% of the nearly 400 identified molecules of the human skin metabolome/volatilome, which includes all verified and well-characterized mosquito attractants. While native skin bacteria are known to be difficult to manipulate genetically, we have recently introduced two new genetic systems in several commensal *Corynebacterium* species, including *C. accolens* and *C. striatum*, as well as multiple primary human isolates of *Staphylococcus epidermidis* without prior lab domestication. By modifying the most common members of the human skin microbiome, we have inserted >10 attractant degradation modules within these strains, demonstrated the long-term engraftment and colonization of synthetic communities comprised of engineered isolates on a preclinical mouse model, and assessed their effects on mosquito repellency through a combination of high-throughput mosquito behavioral assays, metabolomics studies using gas and liquid chromatography mass spectrometry, and species-level microbiome profiling.

Engineered Calprotectin Sensing Probiotics for IBD Surveillance in Humans

Arthur Prindle

Inflammatory bowel disease (IBD) is a spectrum of autoimmune diseases affecting the gastrointestinal tract characterized by a relapsing and remitting course of gut mucosal inflammation. Disease flares can be difficult to predict and the current practice of IBD disease activity surveillance through endoscopy is invasive, time-consuming, and costly. Recent advancements in synthetic biology raise the possibility that symbiotic microbes can be engineered to selectively detect disease biomarkers used in current clinical practice. Here, we introduce an engineered probiotic capable of detecting the clinical gold-standard IBD biomarker, calprotectin, with high sensitivity and specificity in IBD patients. Specifically, we identified a bacterial promoter in the probiotic strain E. coli Nissle 1917 (EcN) which exhibits a specific expression increase in the presence of calprotectin. Using murine models of colitis, we show that reporter signal is activated in vivo during transit of the GI tract following oral delivery. Furthermore, our engineered probiotic can successfully discriminate human patients with active IBD from those in remission and without IBD using patient stool samples, where the intensity of reporter signal quantitatively tracks with laboratory measured levels of calprotectin. Our results demonstrate that probiotics can be engineered to detect fecal calprotectin for precise non-invasive disease activity monitoring in IBD patients.

AI-driven protein discovery for functional ingredients

Ranjani Varadan

Shiru is pioneering AI-driven protein discovery to create functional, sustainable ingredients for the plant-based industry. At the heart of Shiru is Flourish™, a first-of-its-kind platform that combines machine learning, traditional bioinformatics and high throughput wetlab workflows to predict, screen and identify naturally occurring proteins for use as high performance ingredients. The use of cutting-edge ML tools allows Shiru to bypass traditional/empirical trial-and-error based discovery of functional proteins. As a B2B company, Shiru's first product is OleoPro™, a structured fat borne out of Flourish™'s insights into gelation. Shiru is also developing protein replacements for methylcellulose and eggs.

Computer-Aided Design and Implementation of Efficient Biosynthetic Pathways to Produce High Added-Value Products Derived from Tyrosine in *Escherichia coli*

Sofia Ferreira, Alexandra Balola, Isabel Rocha

The aromatic nature of the Shikimate pathway holds a huge commercial interest due to its potential for replacing fossil fuel-derived metabolites[1]. However, the lack of established pathways and known enzymes/genes impairs the development of cell factories. We have established a computational workflow combining retrobiosynthesis algorithms (ShikiAtlas[2]), an enzyme selection tool (BridgIT[3]) and a gene-finding pipeline for efficient pathway design for tyrosine-derived compounds in *Escherichia coli*, specifically L-DOPA, Dopamine and Octopamine.

The pathways for the target metabolites were generated in ShikiAtlas with tyrosine as the precursor. We have considered pathways containing known enzymes or orphan reactions, favoring the maximum carbon conservation ratio and the minimal size. BridgIT was used to find the best enzymes and an in-house structure-based gene discovery pipeline ranked the most suitable sequences. Then, the pathways were implemented using molecular biology methods.

We were able to successfully implement pathways to produce L-DOPA and Dopamine with the most promising known enzymes. Moreover, we have discovered a novel pathway to produce Dopamine avoiding the use of L-DOPA as an intermediate compound. Finally, for the first time, we implemented a biosynthetic pathway to yield Octopamine in *E. coli*. These results validate the computer-aided design of biosynthetic routes to shikimate-derived compounds, encouraging the expansion of this workflow to other aromatics.

1. Averesch, N. J. H. & Krömer, J. O. Metabolic Engineering of the Shikimate Pathway for Production of Aromatics and Derived Compounds — Present and Future Strain Construction Strategies. **6**, 6:32 (2018).
2. Sveshnikova, A., MohammadiPeyhani, H. & Hatzimanikatis, V. ARBRE: Computational resource to predict pathways towards industrially important aromatic compounds. *Metab. Eng.* **72**, 259–274 (2022).
3. Hadadi, N., MohammadiPeyhani, H., Miskovic, L., Seijo, M. & Hatzimanikatis, V. Enzyme annotation for orphan and novel reactions using knowledge of substrate reactive sites. *Proc. Natl. Acad. Sci. U. S. A.* **116**, 7298–7307 (2019).

Systematic Mining of Peptidic Antimicrobials with Diverse Activity Spectra from the Genomes of Human-Associated Bacteria

Zhengan Zhang, Andrew King, Christopher A. Voigt

Human-associated bacteria secrete modified peptides to control host physiology and remodel the microbiota species composition. Here, we scanned the genomes of bacteria colonizing human skin, gastrointestinal tract, urogenital tract, mouth and trachea for gene clusters encoding RiPPs (ribosomally-produced and post-translationally modified peptides). We found 218 lanthipeptides and 25 lasso peptides, 70 of which were synthesized and expressed in *E. coli* and 24 was successfully purified and functionally characterized. They were tested for activity against bacteria associated with healthy human flora and invasive pathogens. New antibiotics were identified against strains implicated in skin, nasal and vaginal dysbiosis as well as from oral strains selectively targeting those in the gut. Broad- and narrow-spectrum antibiotics were found against methicillin-resistant *S. aureus* (MRSA) and vancomycin-resistant *Enterococci* (VRE). Mining natural products produced by microbes living with the human body will enable the elucidation of complex ecological relationships and may be a rich resource for antimicrobial discovery.

Everything as code

David Van Valen

Biological systems are difficult to study because they consist of thousands of parts, vary in space and time, and their fundamental unit—the cell—displays remarkable variation in its behavior. These challenges have spurred the development of genomics and imaging technologies over the past 30 years that have revolutionized our ability to capture information about biological systems in the form of images. Excitingly, these advances are poised to place the microscope back at the center of the modern biologist's toolkit. Because we can now access temporal, spatial, and "parts list" variation via imaging, images have the potential to be a standard data type for biology.

Biology needs a new data infrastructure for this vision to become a reality. Imaging methods are of little use if it is too difficult to convert the resulting data into quantitative, interpretable information. New deep learning methods are essential to the reliable interpretation of imaging data. These methods differ from conventional algorithms in that they learn how to perform tasks from labeled data; they have demonstrated immense promise, but they are challenging to use in practice. The expansive training data required to power them are sorely lacking, as are easy-to-use software tools for creating and deploying new models. Solving these challenges through open software is a key goal of the Van Valen lab. In this talk, I describe DeepCell, a collection of software tools that meet the data, model, and deployment challenges associated with deep learning. These include tools for distributed labeling of biological imaging data, a collection of modern deep learning architectures tailored for biological-image analysis tasks, and cloud-native software for making deep learning methods accessible to the broader life science community. I discuss how we have used DeepCell to label large-scale imaging datasets to power deep learning methods that achieve human-level performance and enable new experimental designs for imaging-based experiments.

AI/ML for Synthetic Biology

Huimin Zhao

Synthetic biology has been rapidly growing in the past two decades. However, due to the complexity of biological systems, performing synthetic biology in a quantitative and predictive manner still remains a challenge. In recent years, thanks to the increasing availability of big data and recent advances in data science, artificial intelligence (AI) and machine learning (ML) that allow computers to learn from experience has emerged as a potentially powerful tool to address this challenge. In my talk, I will highlight our recent efforts in developing and applying AI/ML tools for synthetic biology, including but are not limited to: (a) development of an AI/ML enabled fully closed design-build-test-learn loop named BioAutomata for pathway optimization (1); (b) development of a deep learning model named ECNet for protein engineering (2); (c) development of an AI/ML algorithm named CLEAN (contrastive learning enabled enzyme annotation) to assign enzyme classification numbers to enzymes with better accuracy, reliability, and sensitivity than the state-of-the-art tool BLASTp (3). These AI/ML tools should greatly accelerate the development and application of synthetic biology for a wide variety of biomedical and biotechnological applications.

1) HamediRad, R. Chao, S. Weisberg, J. Lian, S. Sinha, and H. Zhao. "Towards a Fully-Automated Algorithm-Driven Platform for Biosystems Design." *Nature Communications*, 10:5150 (2019).

2) Z Luo, G. Jiang, T. Yu, Y. Liu, L. Vo, H. Ding, Y. Su, W. W. Qian, H. Zhao, and J. Peng. "ECNet is an Evolutionary Context-integrated Deep Learning Framework for Protein Engineering." *Nature Communications*, 12:5743 (2021).

3) Yu, H. Cui, J. Li, Y. Luo, G. Jiang, and H. Zhao. "Enzyme Function Prediction using Contrastive Learning." *Science*, in press.

Deep Learning Sequence Design for mRNA Therapeutics and Human Cell Type-Specific Expression

Sebastian M. Castillo-Hair, Stephen Fedak, Ban Wang, Christopher Yin, Johannes Linder, Kyle Havens, Michael Certo, Wouter Meuleman, Georg Seelig

Engineering biological sequences with desired performance is a major goal of Synthetic Biology. Here, we design *de novo* regulatory sequences for two synthetic biology applications using deep learning predictive models, gradient descent sequence optimization, and generative neural networks. First, we design 5'UTRs to increase performance of a gene editing mRNA therapy. Our strategy is centered around Optimus 5-Prime, a Convolutional Neural Network (CNN) model we previously developed to predict translation efficiency from 5'UTR sequence. After showing that predictive accuracy remains high in cell types relevant for mRNA therapeutics, we designed *de novo* 5'UTRs and experimentally showed that they support strong mRNA-delivered gene editing activity. To design more compact 5'UTRs, we characterized a novel reporter library with >200,000 short, fully randomized 5'UTRs, and trained a new CNN predictor on these data. Finally, we used this model to design a new set of 5'UTRs, and found one that outperforms all designs and controls for one of two tested gene editors. In our second application, we design enhancers to restrict gene expression to specific human cell types. We developed CNN models to predict expression in HepG2 (liver) and K562 (lymphoma) cells using published measurements from >400,000 candidate enhancers. We then used these models to design >1,000 enhancers, and experimentally show that they enable HepG2- or K562-specific expression. To support a larger number of cell types, we developed models of genomic accessibility in 64 cell types using data from >3 million DNase-hypersensitive sites from the ENCODE and Roadmap Epigenetics projects. We then designed candidate enhancers for expression specific to cell types such as SK-N-SH (neural), GM12878 (lymphoid), MCF7 (breast), and H9-hESC (embryonic), which will be experimentally validated. Our results show the potential of deep learning-enabled sequence design, which can be used to test our understanding of cell regulation and to design improved therapeutics.

Engineered Living Materials Programmed by Life

Chao Zhong

Email: chao.zhong@siat.ac.cn

Many biological material systems—such as skeletal tissues, marine adhesives and biofilms—grow, self-repair and adapt to the environment and possess distinctive 'living' attributes that are beyond the reach of the vast majority of existing synthetic materials. Harnessing these attributes to create dynamic, environmentally responsive, and tunable "living" materials is of keen interest in the emerging field of Materials Synthetic Biology. In this talk, I will introduce our recent efforts in creating a highly flexible and tunable engineered living materials (ELMs) platform based on bacterial biofilms by leveraging the power of synthetic biology. The programmable bacterial biofilms, either functionalized with protein modules or regulated by synthetic gene circuits, enable a radically different paradigm of materials synthesis and performance, and provide new opportunities for constructing materials with dynamic features and enhanced sustainability. Several application scenarios including bioremediation, artificial photosynthesis and biomedicine with these living biofilm materials will be highlighted. Finally, I will discuss how multi-disciplinary technologies such as bioinformatics, structural biology and synthetic biology can be integrated to push the discovery and design of new ELMs beyond bacterial biofilms.

Programming Living Materials with Synthetic Polymers and Engineered Bacteria

Seunghyun Sim

In this presentation, I will discuss the Sim lab's recent efforts in developing engineered living materials with a well-defined molecular interface. Their construction (ACS AMI 2022 and in preparation), dynamic properties (JACS 2023), and incorporation of engineered spores for catalytic activities (ACS Syn Biol 2022 and in preparation) will be introduced.

Thinking Soils: Towards Pressure Responsive Microbial Induced Calcite Producing Engineered Living Material

Martyn Dade-Robertson, Jennifer Wright, Polly Moreland, Jamie Haystead, Katie Gilmour, Thora Arnardottir, Mingpei Li, Paul James, Anil Wipat, Meng Zhang

Microbially Induced Calcite Precipitation (MICP) involves triggering mineral crystal formation through enzymatic reactions. In the most used process, the enzymes such as urease and carbonic anhydrase can catalyse the conversion of substrates (urea and CO^2 respectively) to produce carbonate ions. These reactions also cause a pH increase outside the cells which triggers the precipitation of calcite. Bacterial cells also play an essential role in providing a nucleation site for the growth of crystals. MICP has been extensively explored for ground improvement, self-healing concrete and bio cementation.

Our starting point is to engineer the MICP processes to perform more like bone formation, where the production of calcium phosphate is mediated by living cells where structure is remodelled in response to mechanical loads. We envisage a system capable of producing a self-constructing building foundations in which pressure sensing microbes are injected into soils and, as a building is constructed above, the microbes produce calcites, in response to the load, through MICP. The calcite will bind the aggregate together and strengthen the soil. Our approach is multi-scale by 1) identifying a genetic response to elevated pressures in *Bacillus subtilis*, and engineering genes in the cells to be activated by load; 2) engineering *B. subtilis* with controllable urease expression; 3) optimising bacteria activities and behaviours in soils and 3D soil-like matrices, 4) developing computer models which map between geotechnical conditions and genetic response to soils under pressure to predict the outcomes. We report on the latest findings of this study including some of our biological insights, initial results from controllable urease expression and pressure response system. We also address some of the challenges of developing this novel type of large-scale soil based Engineered Living Material and suggest it as the basis for broader applications.

Pathway engineering using recombination-based expression optimization in yeast

Charlotte Cautereels, Anna Zimmerman, Jan Steensels, Anton Gorkovskiy, Kevin Verstrepen

Metabolic engineering of microbes is crucial for the development of sustainable, industrial production of fuels, medicine, nutrition supplements, antibiotics and high-value chemicals. One major bottleneck in metabolic engineering is expression optimization of the heterologous pathway genes, as suboptimal expression profiles can result in the buildup of toxic intermediates, insufficient pathway flux and metabolic burden. Current rational engineering approaches mostly tackle genes one by one and rely on repetitive design-built-test cycles, which are laborious and expensive, or rely on untargeted approaches (such as random mutagenesis), which can result in undesired side-effects. Therefore, we developed a genetic tool for large-scale in vivo expression diversification across multiple pathway genes simultaneously. The tool uses a set of newly developed orthogonal LoxPsym sites and Cre recombination to shuffle various regulatory elements controlling the expression of a pathway gene, such as promoter parts or terminators. The presented LoxPsym variant set allows to rapidly generate structural variation at up to 16 genomic loci simultaneously without cross-reaction between the different targeted loci. This way, variant pools of yeast with different expression profiles can be generated, in which the expression level of each pathway gene varies over 100-fold. As a proof-of-concept, this expression optimization tool was used to optimize the expression of six pathway genes needed for microbial production of astaxanthin, an important high-value anti-oxidant. Profiling 30 variants with randomized gene regulation gave us new insights into optimized expression profiles and let to the identification of improved strains, ultimately resulting in a two-fold increase in titer compared to a rational design approach.

Characterizing and Engineering Recombinant Lens Proteins Toward Optical Materials

Jelle van der Hilst, Angela M. Belcher, Christopher A. Voigt

GRadient INdex (GRIN) lenses offer several advantages over their mono-index counterparts, including total freedom in surface geometry and a lack of spherical aberration. However, their use is limited due to the cost, speed, and poor scaling of current manufacturing techniques. Proteins such as S-crystallins, which self-organize into high-refractive-index (n) gels *in vivo*, show promise for building biological GRIN lenses. Here we present a set of strategies for engineering high-n protein-based materials. We first characterized a set of known and unknown highly refractive proteins, including published cephalopod S-crystallins and the calculated top 50 highest-n sequences from the UniProt database. Separately, we used computational mutagenesis with PyRosetta to predict mutations for raising the refractive index of hypersoluble scaffolds such as bovine serum albumin (BSA). Finally, we set out to build materials with spatially controlled refractive indices, including a poly(acrylamide) hydrogel with a Δn of 0.016, representing a spatial gradient of BSA concentration from 0 to 60 mg/mL. We next look to improve the spatial resolution and maximum n of these materials. This work presents progress toward protein-based optical materials with fully tailorable refractive indices and compositions, enabling cheaper and highly scalable GRIN lens production.

Using CRISPR Cas Editing to Improve Microbial Establishment in the Rhizosphere

Ilenne Del Valle, William Alexander, Joshua Michener, Carrie Eckert

One approach to enhance plant growth and pest resistance is to engineer its rhizosphere by introducing Plant Growth Promoting Bacteria (PGPR). PGPR can improve plant health and productivity by secreting beneficial secondary metabolites and inducing systemic resistance. However, non-native PGPR often fails to establish and survive under different environmental conditions and when competing against the rhizosphere microbiome. The successful establishment of non-native microbe relies on the interaction of multiple phenotypic traits (e.g., biofilm formation, sporulation, and gene transfer) and needs to be better understood. Therefore, we need to study these genetically encoded traits and their interactions on a genome-wide scale.

Our work focuses on *Bacillus velezensis*, a known clade of PGPR and antagonist of the pathogenic fungus *Sphaerulina musiva* that infects the biofuel feedstock *Populus*. We plan to use high-throughput CRISPR-mediated genome editing to obtain genome-wide knockout/knockdown libraries in *B. velezensis*. We can study how the loss of a gene function influences *B. velezensis* establishment in the rhizosphere by testing these knockout/down libraries under different selective conditions.

To obtain sufficient gRNA library coverage, we improved electro-transformation efficiency in this non-model organism by engineering *E. coli* to express *B. velezensis*-specific methylases. We improved transformation up to 10^6 CFU per 1μg of DNA by using DNA with the proper methylation signature. We also characterized different genetic parts to engineer a robust Cas9/dCas9 expression system and built a plasmid-based gRNA genome-wide library targeting ten gRNA per gene. We plan to perform growth-based selection and measure gRNA enrichment with NGS. The data collected from these experiments will validate a gRNA efficiency model. Further growth selection assays and functional analyses will improve our understanding of the genetic factors affecting microbial establishment, persistence, and functionality. This information will help predict the effect of PGPR on plant health and manipulate rhizosphere establishment.

Evaluating CRISPR-Assisted Natural Transformation of *Vibrio Natriegens* at Putatively Neutral Genomic Sites

Eric Vanarsdale, Tanya Tschirhart, Gary Vora

Vibrio natriegens is an emerging host-microbe for biotechnology development in metabolic engineering and synthetic biology applications. Genomic engineering of *V. natriegens* for these applications are assisted by the natural competence of *Vibrionaccae*, many of which natively integrate genetic components of non-kin bacteria. To assist these efforts, our goal in this project is create a predictive paradigm for genomic modification of *V. natrigens* for rapid strain evaluation and development. Here we investigate the capability of a recent CRISPR-assisted selection method, named Natural Transformation CRISPR (NT-CRISPR) to integrate DNA at 7 selected sites from a set of 39 putatively neutral sites identified by Bernhards, C. et al. We have found that the growth rate prior to and after induction of spCas9 correlates to the efficiency of tDNA insertion in PCR screened colonies. Of the 7 sites tested, 5 sites were successfully edited with 1 site causing substantial fitness loss to the cell. When comparing these insertions for inducible expression of GFP, we have observed expression from constructs integrated on chromosome 1 result in greater expression during the mid and late log growth phases than those located on chromosome 2. Constitutive constructs did not consistently show the same trends. Lastly, we discuss future work to further development of genomic integration platforms.

Genome-Wide Crispri Screening Identifies Enhanced Autolithotrophic Phenotypes in Acetogenic Bacteria

Jiyun Bae, Jongoh Shin, Hyeonsik Lee, Seulgi Kang, Sangrak Jin, Minjeong Kang, Eojin Yoo, Yoojeong Choi, Suhyung Cho, Ki Jun Jeong, Byung-Kwan Cho

Acetogenic bacteria are a unique biocatalyst that highly promises to develop the sustainable bioconversion of carbon oxides (e.g., CO and CO_2) into multi-carbon biochemicals. Genotype-phenotype relationships are important for engineering their metabolic capability to enhance their biocatalytic performance, however, systemic investigation on the fitness contribution of individual gene has been limited. Here, we report genome-scale CRISPRi screening using 41,939 guide RNAs designed from the *Eubacterium limosum* genome, one of model acetogenic species, where all genes were targeted for transcriptional suppression. We investigated the fitness contributions of 96% of the total genes identified, revealing the gene fitness and essentiality for heterotrophic and autotrophic metabolisms. Our data show that the Wood-Ljungdahl pathway, membrane regeneration, membrane protein biosynthesis, and butyrate synthesis are essential for autotrophic acetogenesis in *E. limosum*. Furthermore, we discovered genes that are repression targets that unbiasedly increased autotrophic growth rates 4-fold and acetoin production 1.5-fold compared to the wild-type strain under CO_2-H_2 conditions. These results provide insight for understanding acetogenic metabolism and genome engineering in acetogenic bacteria. This work was supported by the C1 Gas Refinery Program (2018M3D3A1A01055733 to B.-K.C.), the National Research Foundation of Korea grant (2021R1A2C1012589 to S.C.), and the Bio and Medical Technology Development Program (2021M3A9I4024308 to B.-K.C.) through the National Research Foundation of Korea funded by the Ministry of Science and ICT.

Development of CO Gas Conversion System Using High CO Tolerance Biocatalyst

Seulgi Kang, Sangrak Jin, Jiyun Bae, Hyeonsik Lee, Minjeong Kang, Eojin Yoo, Byung-Kwan Cho

Acetogenic bacteria (acetogens) are microorganisms with high potential as biocatalysts that can convert C1 gas, such as carbon dioxide (CO_2) or carbon monoxide (CO), into biomass or biochemicals; however, high CO concentrations cause growth retardation in these microorganisms. In this study, we performed adaptive evolution of *Eubacterium limosum*, an acetogen, for approximately 400 generations under high CO synthetic gas (syngas) conditions to obtain *E. limosum* ECO2, which can rapidly proliferate even under high CO conditions. A mutation was revealed in H636R in the Acetyl-CoA synthase (ACS) protein of the ECO2 strain through whole-genome re-sequencing, and the mutation increased the CO gas conversion rate. By analyzing the transcript profiling of the ECO2 strain under high CO syngas conditions, the transcript levels of the methyl branch in the Wood–Ljungdahl pathway (WLP) and the energy conservation system increased at high CO concentration conditions. Finally, 2,3-BDO (2,3-butanediol) biosynthesis pathways were constructed and transformed into the *E. limosum* ECO2 strain to convert CO into value-added biochemicals. This strain showed the capability to produce 2.60 mg $L^{-1}h^{-1}$ of 2,3-BDO through CO 66% syngas fed-batch fermentation, which was 6.5-fold higher than the wild type. We developed a microorganism with high CO tolerance, and this strain showed high application potential as a biocatalyst that can efficiently convert CO gas into value-added chemicals. This work was supported by the C1 Gas Refinery Program (2018M3D3A1A01055733 to B.-K.C.) through the National Research Foundation of Korea funded by the Ministry of Science and ICT.

Engineering Production of Specialty Fragrant C15 Sesquiterpenoids from the Green Algal *Chlamydomonas Reinhardtii*

Sergio Gutierrez, Sebastian Overmans, Gordon Wellman, Kyle Lauersen

Terpenoids, a family of natural compounds with various uses, are naturally produced by many organisms, including plants, bacteria, and mammals. However, some of these specialty chemicals are currently being sourced in an unsustainable manner from slow-growing plants in sensitive environments. One notable example of this is the fragrant products of the *Aquilaria* tree; the resinous wood is used in the Middle East as steam distillate called 'oud,' and the dried wood chips are burned as incense called 'bakhour.' Heterologous production of terpenoids in engineered microbes provides an alternative source for these chemicals. Recent research has leveraged metabolic engineering in the photosynthetic model green microalga, *Chlamydomonas reinhardtii*, as an alternative cell chassis to produce heterologous terpenoids. This organism is interesting because it consumes CO_2 for its carbon and light as an energy source through photosynthesis. The research presented here aims to produce various heterologous sesquiterpenoids in *C. reinhardtii* and enhance flux to these products by modifying its native pathway to sesquiterpenoids. The host squalene synthase (SQS) was knocked down, as previously reported, to decrease the metabolic pull of farnesyl pyrophosphate (FPP), and a beta-carotene ketolase (BKT) was induced to perturb carotenoid biosynthesis. Surprisingly, these modifications synergized FPP availability, increasing patchoulol production 30-fold (5.6 ± 1.4 fg/cell to 183 ± 23 fg/cell). Using a strain with a perturbed carotenoid pathway and reduced SQS expression, producing fragrant terpenoid chemicals often found in *Aquilaria* trees was undertaken and benchmarked *in alga*. Various terpene synthases to produce aristolochene (n=4), guaiene (n=6), santalene (n=3), valencene (n=1), valerianol (n=1), zizaene (n=2), cadinol (n=1), and bisabolol (n=1) were screened for activity directly in *C. reinhardtii* using optimized synthetic transgene designs. We show that the alga can make all these non-native terpenoids and may be an alternative source of fragrant specialty chemicalsderived sustainably from CO_2.

Synthesis of Fatty Acid Esters of Hydroxy Fatty Acids

Lay Hiang Ling

Lipases are ubiquitous in nature and are frequently used in industrial applications as they can catalyse various reactions. Recently, Enz1 lipase was discovered to be able to catalyse the synthesis of FAHFA (fatty acid esters of hydroxy fatty acids), which is a bioactive molecule that has potential to be developed for the treatment of diabetes and inflammatory diseases. To expand the catalogue of enzymes available for the synthesis of various types of FAHFA, and to facilitate functional studies, a sequence similarity network was constructed for Enz1. Orthologues of Enz1 were synthesized and expressed in a GRAS (generally recognized as safe) chassis, *Pichia pastoris*, and were subsequently screened and characterized. Out of the 57 orthologues, 38 demonstrated activity and 17 could accept long chain triglycerides. These orthologues were also found to be thermostable. Notably, four of the orthologues exhibited higher expression than Enz1, and two of these were able to synthesize 12-PAHSA (12-palmitate-hydroxystearic acid) from 12-HSA (12-hydroxystearic acid) and PA (palmitic acid). One of the orthologues was successfully crystallized and was found to be different from the structure predicted by AlphaFold2. Future studies will focus on determining the rate of FAHFA synthesis by the identified enzymes, exploring their substrate specificity, and conducting functional studies on the synthesized FAHFA.

Design and Evolution of Rhodopsin-Based Fluorescent Biosensors for Chloride

Shelby Phelps, Jasmine Tutol, Qin Zhou, Hsichuan Chi, Jessica Lee, Deeya Advani, Sameera Abeyrathna, Paarth Kapadia, Gabriele Meloni, Faruck Morcos, Sheel Dodani

Chloride is a vital anion for life. Our understanding of chloride in biology has been accelerated by fluorescent protein-based sensors derived from the green fluorescent protein (GFP) family. To expand beyond these advances, we envisioned that members of the microbial rhodopsin family could also serve as a scaffold to engineer fluorescent biosensors for chloride. Microbial rhodopsins are proteins with red absorption and emission spectral profiles that inherently localize to the membrane. These natural properties have been enriched via protein engineering methods for bioelectronic and optogenetic applications. Building from this knowledge, we have uncovered that chloride sensitivity can be unlocked via a single mutation of the residue that serves as the counterion to the protonated Schiff-base chromophore in the proton-pumping rhodopsins *Gloeobacter violaceus* and *Halorubrum sodomense*. This has resulted in three generations of chloride biosensors: GR1, GR2, and ChloRED-1. In this presentation, I will discuss the design, evolution, spectroscopic properties, and bacterial imaging applications of this biosensor family.

Establish the Polyextremophilic Red Microalga *Cyanidioschyzon Merolae* As a Host for Synthetic Biology and Scalable Production of Valuable Biomass

Melany Villegas, Barbara Bastos De Freitas, Fatimah Abualsaud, Martha Stark, Stephen Rader, Gabriel Ivan Romero Villegas, Ricardo Gonzalez, Claudio Grunewald, Kyle Lauersen

The Cyanidiales are a group of unicellular red algae that thrive in thermo-acidic environments (pH 0.05-5.0, up to 60 °C) and environments with elevated heavy metal concentrations. *Cyanidioschyzon merolae* 10D is found in acid hydrothermal systems and it can be cultivated in acidified medium (pH 0.5-3) to high biomass in both lab-scale and outdoor conditions with minimal contamination. This alga may be a powerful candidate for synthetic biology and metabolic engineering for heterologous product generation due to its simple genomic features – a small nuclear genome (16.5 Mb) with few introns (0.5%) and 55% GC content– Moreover, unlike other algae, nuclear transformation and integration of transgenes into chromosomes can occur by homologous recombination. Although molecular tools for the successful transformation of this alga have been recently developed, current methods to generate transformants require weeks to months. Here, we report an optimized transformation and selection protocol for *C. merolae* 10D, generating single chloramphenicol resistant colonies within 6-8 days. We present an *in silico* designed and *de novo* constructed modular plasmid toolkit, its transformation, and benchmark transgene expression by fluorescence imaging, flow cytometry, in-gel fluorescence and Western blotting. The transformation and stable expression of several fluorescent reporters in both linear DNA fragments and circular episomes, as well as the application of metabolic engineering for the production of volatile isoprene in this alga is presented for the first time in this work. Additionally, we demonstrated the use of *C. merolae* as a system for large-scale cultivation with continuous CO_2 injection to produce ~2 g L^{-1} of valuable biomass in 1000 L tubular photobioreactors in the extreme environment of the mid-Red Sea coast during August 2022. This work aims to show that *C. merolae* is an emerging algal system for metabolic engineering and synthetic biology mediated sustainable and scalable production of engineered bio-products.

Synthetic RNA Interference Agents for Species-Specific Invasive Plant Control

Ping Gong, Seung Ho Chung, Kurt Kowalski

RNA interference (RNAi) is a key mechanism for gene expression regulation at the post-transcriptional level. Many synthetic biology technologies have been developed for pest control and gene therapies by harnessing the natural RNAi process. However, application of RNAi-mediated gene silencing to invasive plants control has not been adequately explored. As a proof-of-concept study, we designed and tested multiple classes of synthetic RNAi agents to knock down essential genes in target invasive plants. These RNAi agents ranged from artificial microRNA and double-stranded RNA to chemically modified and non-modified antisense oligonucleotide, as well as engineered plasmids and viral vectors. We also screened a variety of nanoparticles as potential carriers, including cell penetrating peptides, carbon nanodots, carbon nanotubes, and clay nanosheets, to enhance the delivery of synthetic RNAi agents and efficiency of interruption or suppression of gene expression. We will use the invasive common reed *Phragmites australis* as an example to introduce the basic concepts and principles of RNAi-based invasive species control, demonstrate recent technological developments, and discuss existing technical barriers and perspective on the future of RNAi-based next-generation bioherbicides.

Enhanced Genome-Wide Knockout Screens in Bacteria with CRISPR Base Editors

Sandra Gawlitt, Scott Collins, Yanying Yu, Samuel Blackman, Lars Barquist, Chase L. Beisel

Genome-wide knockout or knockdown screens have become powerful tools for the investigation of genotype-to-phenotype relationships. In bacteria, these screens commonly rely on transcriptional repression with catalytically dead (d)Cas9 or gene knockouts through Cas9 editing or random transposon mutagenesis. However, each approach comes with a notable limitation such as incomplete gene silencing or massive libraries. Here, we take a distinct approach to genome-wide screens in bacteria with base editing. We use the base editors to introduce premature stop codons or mutate start codons using a ScCas9 nickase derived base editor (ScBE3) that exhibits flexible PAM recognition. We then derive guide design rules by applying machine learning to a gene essentiality screen conducted in *Escherichia coli*. To address instances of poor editing, we combined base editing with Cas9-induced cleavage of the unedited cell fraction, thereby enriching edited cells. The efficiency of this dual system was validated through a screen of conditionally essential *E. coli* genes, which yielded more efficient guide depletion and confirmed previously published conditionally essential genes. Overall, base editing represents a useful tool for genome-wide knockout screens in bacteria and will eventually enable genome-wide knockout screens in a broader range of bacterial species.

Growing Convergence Research from the Lens of Biomolecular Systems Engineering

Ishita Kumar

The National Science Foundation recently identified Growing Convergence Research as one of the 10 'Big Ideas' that will be essential for developing the fields of science and engineering in the coming years. Convergence research is motivated from the understanding that solving today's global problems requires multidisciplinary expertise and a platform to create novel scientific frameworks. The Wilson Lab recognizes the potential of employing convergence research via developing the field of Biomolecular Systems Engineering (NSF Award # 1934836). Our lab works towards combining principles from biological, electrical, and chemical engineering to establish biological programming in a range of chassis cells. This strategy directly addresses the concerns of a lack of systemization in biological engineering by redefining the workflows - i.e. taking inspiration from process design, and building application-agnostic biological unit operations. Biological programming with this approach can be utilized to create the foundation for modular and efficient drug delivery, cell manufacturing, and biosecurity systems. To accomplish the aforementioned, requires the development of intelligent biological chassis cells, which are imbued with synthetic (i) decision-making, (ii) memory, and (iii) communication. We believe that our novel and systematic workflows, emerging from convergence research, will produce bespoke systems for use in a myriad of applications.

Microbial Community Transcriptomics: Practices and Challenges

Jeanet Mante, Randi Pullen, Bryn Adams

Understanding of natural microbial communities and their interactions within the community and with their environment is critical for the exploration of natural resources for a variety of biotechnological applications. This requires the use of transcriptomics, which is the study of RNA transcripts produced by a specific organism. Community transcriptomics methods and techniques are still being developed. These efforts seek to optimize the process of analyzing RNA from complex environmental samples, which will enable a deeper understanding of the potential applications of bioprospecting. We offer an overview of current practices and the remaining challenges.

Developing High-Throughput Dbtl Cycle for Pks Based Retrobiosynthesis

Namil Lee, Matthias Schmidt, Chenyi Li, Sarah Klass, Anna Lisa Fear, Tyler Backman, Christopher J. Petzold, Robert W. Haushalter, Jay Keasling

Microbial cells have been engineered to produce several commodity bioproducts and biofuels. However, the range of products has been limited by the metabolic pathways that can be engineered into cells and the lack of engineered enzymes, particularly those that can form carbon-carbon bonds. Polyketide synthases (PKSs) form carbon-carbon bonds and have the ability to produce a nearly infinite number of organic molecules. The modularity of PKSs render them excellent targets for engineering (at the DNA level) to produce existing and novel small molecules. Given their potential, PKSs are underutilized for biological synthesis of organic molecules because of the challenges in programming them, in part due to the lack of an adequate Design-Build-Test-Learn (DBTL) cycle to assemble and express PKSs in a suitable host and produce the desired final product. To this end, we developed and combined software to design PKSs, automated (robotic) pipeline to build PKS genes, proteomics and metabolic analysis to test PKS expressing host, and machine learning tools to learn and improve future PKS designs. The combination, and enhancement, of these tools will create a game-changing technology platform for engineering PKSs, which will greatly benefit the industrial biological production of chemicals and fuels from cellulosic biomass.

Five-Aminovaleric Acid and Glutaric Acid Production in *Novosphingobium Aromaticivorans* from Lignin-Based Aromatics

Christopher Azubuike, Joshua Michener

Lignin is abundant source of aromatic polymers, but it is underutilized due to the complexity and variability of its mixture following depolymerization. *Novosphingobium aromaticivorans* can use a range of lignin-derived aromatic substrates for growth. We focused on this bacterium central metabolism intermediates (pyruvate and oxaloacetate), through lysine biosynthetic pathway, to accumulate 5-aminovaleric acid (5-AVA) and glutaric acid (GTA). These two compounds are important platform chemicals. While evolution to improve lysine biosynthesis for conversion to 5-AVA and GTA so far did not increase production, discovery metabolomics revealed overproduction of amino acids and their metabolites. The production of lysine as the precursor of 5-AVA and GTA in *N. aromaticivorans* appears to be tightly regulated. Identifying the genetic target(s) to derepress lysine biosynthesis, and subsequently deleting 5-AVA and GTA catabolic pathways will improve production. Additionally, amino acids overproduction in *N. aromaticivorans* evolved strains can be leveraged to produce other important biochemicals.

Engineering Bacteria for the Biodegradation and Upcycling of Nylon Monomers

Mirna Kheir Gouda

We are engineering bacterial strains for upcycling of HMDA, 6-ACA and adipic acid which are the biodegradation products of nylon 6 and nylon 6,6. A single four-step enzymatic pathway has been designed to encompass all three nylon monomers where HMDA and 6-ACA are converted to adipic acid which metabolizable through beta-oxidation in some microbes. Notably, *Acinetobacter Baylyi ADP1* is capable of utilizing adipic acid as the sole carbon source for growth. Heterologous enzymes were tested for promiscuous activity and active enzymes along with importer proteins were assembled into a genomically expressed operon. Additionally, competing native metabolic reactions were eliminated to maximize metabolism of the monomers. The engineered *A. Baylyi* strain was shown to grow on all three monomers as the sole carbon source. With efficient bioassimilation of nylon monomers, we are expanding the *de novo* pathway's toolkit to enzymatically derive industrially valuable chemicals, such as γ-aminobutyric acid, aminolevulinic acid, ectoine and L-theanine, from pathway-generated metabolites that are non-essential for growth. This coupling of biodegradation and biosynthesis pathways enables microbes to derive energy from waste to generate value-added products and highlights a sustainable synthetic biology-based solution to the global plastic pollution crisis.

Morphogenetic Effects of Mucin Overexpression in Multicellular Spheroids of Mammalian Cells

Zoe Shappell, Leonardo Morsut

All cells possess a glycocalyx; a "shell" of glycoproteins and glycolipids that surrounds the cell membrane. The glycocalyx plays a significant role in cell signaling, adhesion, and survival and changes dynamically with cell state transitions and environmental perturbations. Major components of the mammalian glycocalyx are glycoproteins called mucins. Mucins are composed of a cytoplasmic tail, a transmembrane domain and an extracellular polypeptide backbone that is heavily glycosylated with branching sugar side chains called O-glycans. Since mucins are known to affect cell survival, adhesion, and signaling, synthetic mucins were developed to control these cell biology parameters. However, a characterization of the effects of mucin overexpression on the behavior of mammalian cells in multicellular systems is currently lacking. To study the impact of mucin overexpression on cell adhesion, we set up a live imaging spheroid aggregation assay. Mucin-inducible HEK 293T cells were aggregated with parental HEK 293T cells to form a spheroid. One day after aggregation, mucin was induced with doxycycline. The mucin-expressing cells underwent dramatic morphological changes and began to phase-separate from the spheroid to form a layer around the aggregated parental cells. Large spaces surrounded each mucin-expressing cell, similar to what is shown in vivo as responsible for tissue deformations like embryonic trunk tissue elongation. Mucins could be valuable tools to program morphogenetic rearrangements. We will discuss assays to assess whether mucin overexpression affects cell-cell signaling in multicellular developing systems. This work will lay the foundation for the development of specific circuits for controlling morphogenesis and cell-cell signaling.

Systems Biology of Competency in Bacteria Is Revealed By Applying Novel Data Analytics to the Transcriptome

Jongoh Shin, Kevin Rychel, Bernhard O. Palsson

Vibrio species regulate their natural competency through activation of two transcription factors, TfoX and QstR, involved in the capture of external DNA and its transport into the cell. However, the comprehensive genetic and transcriptional regulatory basis for competency remains unknown. Here, we use a machine-learning approach to decompose *Vibrio natriegens'* transcriptome into 45 groups of independently modulated sets of genes (iModulons). We discovered that competency depends on the repression of two housekeeping iModulons (iron metabolism and translation) and the activation of six iModulons; including TfoX and QstR, a novel iModulon of unknown function, and three housekeeping iModulons (representing motility, polycations, and ROS responses). Phenotypic screening of 83 gene deletion strains across the six iModulons showed that loss of iModulon function reduces or eliminates competency. New data analytics revealed the transcriptomic basis for competency and its relationship to housekeeping functions. These results provide the genetic basis for systems biology of competency.

Dynamic Division of Labour in Synthetic Bacterial Communities for Efficient Bioproduction

Harman Mehta, Rodrigo Ledesma Amaro, Guy-Bart Stan

Recent progress in the field of bioengineering has vastly enhanced our capability of engineering organisms to perform desired functions. While synthetic biology predominantly focuses on engineering single cells to introduce desired functionality, engineering additional functionality into the cell can increase metabolic burden and induce genetic instability, making engineered cells more prone to negative selection. In nature, cells seldom exist in isolation and rather grow in diverse interacting communities. Microbial communities have been shown to demonstrate higher robustness to environmental perturbations and resilience to mutations. The limitations of isogenic populations of engineered cells in monocultures can be mitigated by designing synthetic microbial cocultures performing division of labour, which is the splitting of metabolic pathways across multiple organisms, thus reducing metabolic burden on a single cell. This allows for efficient integration of more complex pathways, provided that the pathways can be efficiently split between cells.

In synthetic microbial consortia, it is important to have a regulatory mechanism in place to control the composition of the coculture. This is because differences in subpopulation growth rates can eventually cause one species to outcompete the other leading to its extinction from the coculture. Due to the challenges in developing control circuits for the dynamic control of coculture compositions, little is known about the effect of continuous ratiometric control on bioproduction. In this study we investigate the advantages of division of labour for the bioproduction of high value compounds, with applications in biomedicine, cosmetics and food. In order to demonstrate the advantage of dynamic control of coculture composition, we develop a synthetic two-member bacterial community splitting two metabolic pathways for bioproduction, the mevalonate pathway and the shikimate pathway. We compare the characteristics of the coculture and the monoculture to identify the suitability of cocultures for bioproduction.

Synthetic Protein Quality Control to Enhance Full-Length Translation in Bacteria

Yong Hee Han, Jina Yang, Jongwon Im, SangWoo Seo

Coupled transcription and translation processes in bacteria cause indiscriminate translation of intact and truncated mRNAs, inevitably generating non-functional polypeptides. Here, we devised a synthetic protein quality control (ProQC) system that enables translation only when both ends of mRNAs are present and followed by circularization based on sequence-specific RNA–RNA hybridization. We demonstrate that the ProQC system dramatically improved the fraction of full-length proteins among all synthesized polypeptides by selectively translating intact mRNA and reducing abortive translation. As a result, full-length protein synthesis increased up to 2.5-fold without changing the transcription or translation efficiency. Furthermore, we applied the ProQC system for 3-hydroxypropionic acid, violacein, and lycopene production by ensuring full-length expression of enzymes in biosynthetic pathways, resulting in 1.6-fold to 2.3-fold greater biochemical production. We believe that our ProQC system can be universally applied to improve not only the quality of recombinant proteins production but also efficiencies of metabolic pathways.

Direct Evolution through Natural Transformation Enables Alginate Assimilation in *Vibrio Natriegens* for Sustainable Bioprocessing

Yungyu Lee, Mincheol Choi, Keonwoo Kim, SangWoo Seo

Alginate is a major component of brown macroalgae, and its efficient utilization is critical for the development of sustainable technologies where abundant marine biomass could be converted into valuable products. *Vibrio natriegens* is a fast-growing marine based bacterium that has gained huge attention in the biotechnological scene due to its potential as an alternative industrial chassis. However, *V. natriegens* cannot naturally metabolize alginate, limiting its usage in marine biomass conversion.

In this study, we integrated the 33kb of alginate assimilation pathway from *Vibrio* sp. dhg into *V. natriegens* through Direct Evolution through Natural Transformation (DENT), resulting in a novel engineered strain that can degrade and consume marine biomass alginate. We further demonstrated the potential of this engineered *V. natriegens* strain for the production of isopentenol, a valuable chemical used in the production of fuels, plastics, and other products. This study presents a new strategy for the metabolic engineering of *V. natriegens* and highlights the potential of this engineered strain for the conversion of marine biomass into valuable products.

Robust Performance and Rapid Construction of Live Bacterial Therapeutics Lacking the Colibactin Gene Cluster

Jaclyn Thompson, Aida Kalantari, Michael James, Lauren Renaud, Mary McDonald, Sean Cotton, Jillian Means, Britta Goncarovs, Vincent Isabella, Mylene Perreault, David L. Hava, Analise Reeves

Synthetic Biotic medicines utilize non-pathogenic probiotic *E. coli* Nissle (EcN) as the chassis organism for engineering strains with drug-like properties designed to perform therapeutic functions in patients. The EcN chromosome in Synthetic Biotics is modified with additional stability features, such as the inclusion of auxotrophy for biocontainment and removal of endogenous phage to prevent unintended replication. However, concerns have been raised regarding the presence of the native *pks* gene cluster, encoding the genotoxin colibactin, prompting evaluation of the effects of removing of the *pks* island from Synthetic Biotics. Here, we report that *Δpks* EcN strains maintained engineered activity and displayed no growth disadvantage when tested in *in vitro* assays. Additionally, there was no significant difference in *in vivo* strain performance between the clinical-stage strain SYNB1934 and its isogenic Δ*pks* variant with regards to recovery of the quantitative strain-specific biomarkers, transit time, and colonization in *in vivo* preclinical mouse and nonhuman primate models. Together, the data demonstrate that *pks* island removal is a well-tolerated chassis modification. Next, to facilitate rapid construction of future Synthetic Biotics, a collection of 24 chassis strains, dubbed the Universal Chassis collection, were assembled, containing each necessary chassis modification, including Δ*pks,* as well as landing pad integration sites for expression of therapeutic modalities. Previously, each modification was added one by one using recombineering with 2-3 weeks needed per edit. In total, around 12-18 weeks were needed to produce clinic-ready prototypes. With the Universal Chassis collection available, all necessary chassis modifications are already in place and modality-specific features can be integrated simultaneously using landing pads. Now, clinic-ready prototypes can be produced in as little as 4-6 weeks.

Engineering *E. coli* for the Utilization of Ethylene Glycol As a Sole Carbon Source for Bioproduction

Michelle Feigis, Yilan Liu, Matthew Edghill, Radhakrishnan Mahadevan

The transition to a low-carbon bioeconomy necessitates the development of novel bio-based processes to produce fuels and chemicals that are decoupled from fossil fuels, sustainable, and economically viable. Most modern-day bioprocesses rely on sugars derived from cellulosic feedstocks that often compete with human consumption and raise concerns about responsible resource management and land use, leading to the emergence of unconventional non-biomass feedstocks as alternatives. While the direct capture and use of carbon dioxide is an appealing option, in vivo carbon fixation has yet to reach industrially relevant levels in engineered microorganisms. Alternatively, intermediate organic compounds can be derived from CO_2 through non-biological means such as electrochemical reduction (ECR) and more readily converted to desired products through metabolic engineering. A compound of particular interest is ethylene glycol (EG), an industrial chemical that is also a common pollutant and degradation product of polyethylene terephthalate plastic. In the presented research, an *Escherichia coli* strain with the ability to consume EG was engineered and became capable of using the substrate as a sole carbon source through adaptive laboratory evolution (ALE). Improved growth rates and a shortened lag phase were also observed after >100 generations. The genetic and transcriptional basis of the adaptations is determined via next-generation sequencing omics analysis. An in-silico comparison of existing and conceptual EG assimilation pathways through metabolic modelling will also be presented. We expect that with further improvements EG can serve as a promising novel bioprocess feed material for a variety of applications.

Assembly of Vaccines Using Cell-Free Integration of Viral Membrane Proteins into Liposomes

Vivian Hu, Neha Kamat

In 2018, the World Health Organization called for accelerated R&D on a list of priority diseases, including the Nipah virus infection, with no efficacious drugs or vaccines that have potential to cause future public health crises. Vaccines remain an effective protective tool against future threats, but many current vaccine strategies and their production techniques have limitations restricting global impact. Cell-free protein synthesis systems (CFPS) are a promising tool for vaccine development because they can rapidly produce clinically-relevant amounts of protein, can be freeze-dried for distribution, and would prevent the need for cold chain storage. Yet, most viral vaccine targets are membrane proteins, which can be difficult to produce with CFPS alone. We and others have shown that adding liposomes to these systems provides a scaffold to promote the cotranslational expression and folding of membrane proteins, resulting in one-step assembly of lipid nanoparticles with membrane-integrated proteins. This technology has the potential to assemble potent viral mimetic vaccines that present viral antigens and adjuvants, however, the design rules for engineering such vaccines remain unknown. Thus, towards the goal of vaccine development, we have investigated how lipid composition affects the assembly of cell-free viral mimetics using the Nipah virus (NiV) infection as a model and the two immunogenic membrane proteins: glycoprotein (NiV G) and fusion (NiV F). We have found that viral membrane proteins can be cotranslationally integrated into liposomes using CFPS and that changing lipid composition enhances NiV F and NiV G association with membranes. Cell-free expressed proteins maintain native folding conformation, as assessed by binding to conformational antibodies for NiV F and G. Lastly, the addition of lipid adjuvant monophosphoryl lipid A (MPLA) does not hinder expression. Our work has the potential to be adapted for other viral membrane proteins to rapidly generate potent vaccines for other infectious diseases.

Use of Mesophilic Prokaryotic Argonautes for Sequence Independent Scarless DNA Assembly

Kasra Tabatabaei, Rita Vaiskunaite, Eric Hunt, Andrew P. Sikkema, Katharina Bilotti, Eric J. Cantor, Gregory Lohman

Modern molecular and synthetic biology rely on the generation and manipulation of large DNA constructs ranging from kB sized genes to 100s of kB sized pathways, small genomes, and even artificial chromosomes. As current direct synthesis technologies do not allow for the direct writing of more than several hundred bases, many *in vitro* and *in vivo* methods have been devised to assemble shorter oligonucleotides into the final large targets. Homology directed methods (e.g. Gibson, NEBuilder) is largely sequence independent, but require long overlaps, have issues with repetitive DNA and short fragments, can introduce mutations at fusion sites due to repeated exo/pol rewriting of the junctions, and are typically restricted to 4-6 fragments per assembly. Ligation directed methods (e.g. Golden Gate assembly), have successfully been used to assemble 35 or more fragments in a single round and can assemble repetitive DNA, but typically require that native Type IIS restriction sites be removed, requiring changes to natural DNA sequences to ensure method compatibility.

To address these limitations, we present progress towards the use of *Clostridium butyricum* argonaute (*Cb*Ago) as a mesophilic, versatile, programmable, DNA-guided assembly tool. We demonstrate that *Cb*Ago can be activated *in vitro* via the DNA strand unwinding activity of a nuclease-deficient mutant of RecBC DNA helicase from *Escherichia coli* ($RecB^{exo-}C$) and efficiently cleave linear double-stranded DNA. We furthermore show that *Cb*Ago can generate arbitrary overhangs of virtually any length or sequence, which can be ligated with compatible overhangs to assemble two sequences efficiently and accurately. Our goal is to develop a platform based on *Cb*Ago-$RecB^{exo-}C$ that enables rapid scarless construction of large DNA molecules from a diverse set of DNA components with varying lengths under mild temperature and buffer conditions without any need for Golden Gate-like sequence modification.

Reproducible Living Biosensors Built By Printing Bacterial Materials

Francesca Usai, Giada Loi, Franca Scocozza, Massimo Bellato, Ignazio Castagliuolo, Michele Conti, Lorenzo Pasotti

The combination of synthetic biology and material science allows the creation of smart materials with unique features using engineered organisms [1]. Engineered living materials (ELMs) have potential applications in various areas, such as diagnostics, bioremediation and environmental monitoring. Here, we developed a workflow to create reproducible bacteria-based ELMs with biosensing capabilities. First, we established that bacteria grow up to two days, and subsequently we observed a decrease due to the waste products and insufficient nutrient supply. However, we demonstrated that the cell viability can be improved by media replacement, and the ELMs can be stored in the fridge up to one month. According to literature, protein expression seen under fluorescence microscope was higher at the surface of the structures probably caused by the different oxygenation level. Quantification of the output can be accomplished in the fluorescence microplate readers by analyzing a piece of the ELM, or for a more user-friendly use taking a photo and exploiting the a*channel of the CIELAB space color. Next, we moved on to the characterization of ELM-based biosensors in optimal and non-optimal incubation, starting from batches of different materials and biological replicates, demonstrating a low variability of the outcome even if printed in different days. The effectiveness of the biosensors in realistic environments has been demonstrated by successful detection in soil, tap water, and clinical samples. However, the potential of these materials can only be fully achieved through efforts to automate their production and establish biocontainment strategies [2]. Overall, these customizable ELMs could represent an important aid to address real-world challenges.

[1] Rodrigo-Navarro, A., et al. (2021). Engineered living biomaterials. *Nature Reviews Materials*, 6(12), 1175-1190.

[2] Tang, T. C., et al. (2021). Materials design by synthetic biology. *Nature Reviews Materials*, 6(4), 332-350.

CRISPR/Cas9 Genome Engineering for Customized Bacterial Outer Membranes

Erin Antoshak, Nathan McDonald

Synthetic biology and genome engineering capabilities have facilitated the utilization of bacteria for myriad of applications. The bacterial outer membrane, specifically the lipopolysaccharide (LPS), plays an integral role in physiology, pathogenesis, and is a main target of existing detection assays. The LPS macromolecule is generally composed of three distinct sections, the inner lipid region known as the Lipid A which links directly to a conserved core oligosaccharide and finally the O-antigen which consists of repeating carbohydrate units. Through advances in synthetic biology, the LPS structure can be targeted via genetic engineering of biosynthesis pathways and tailormade for biotechnological applications. We hypothesized that if an LPS genetic biosynthesis system is transferred from one organism to another, then the recipient will produce the customize LPS structure. Utilizing an *Escherichia coli* strain, synthetic LPS biosynthesis genes from *Yersinia pestis* were inserted into the genome with the goal of an engineered *E. coli* with an LPS like *Y. pestis*. The scaffold *E. coli* strain "ClearColi" has deletions of the native LPS genes and produces the minimal membrane required for survival. Using CRISPR/Cas9 recombineering we first transferred the Lipid A acyltransferase genes *lpxM* and *lpxP*, from *Y. pestis* into our scaffold. Next, we restored the ability to biosynthesize the Lipid/Core linker carbohydrate, Kdo, via insertion of the *Y. pestis kdsD* gene. The Lipid A modifying enzymes EptA and ArnT were added under the control of an inducible promoter. The expression and functionality of each of the replaced genes was confirmed via RT-PCR and MALDI-TOF analyses of isolated Lipid A. Finally, the addition and activation of the core oligosaccharide system is expected to result in a fully assembled *Y. pestis* LPS structure on an *E. coli* backbone. Engineered and designed LPS structures can translate to areas of research including novel therapeutics, diagnostics and detection applications.

Functional Yeast Display of Human Ion Channels That Modulate Pain

Caitlin Tramont, Elizabeth Gardner, Colleen Mulvihill, Raul Trevino, Clayton Kosonocky, Aaron Feller, Andrew D. Ellington

TRPV1 is a non-specific cation channel known for its role in nociception, and as a result, it is a possible target for treating acute and chronic pain. However, its role in the pain pathway has proven to be complex, in part because of its expression across multiple different types of neurons, and its connectivity to multiple different signaling pathways. Current experimental methods heavily rely on mammalian cell expression, which can limit the speed of experimentation and the ability to orthogonally manipulate signaling pathways. Therefore, we have functionally expressed this channel in *Saccharomyces cerevisiae*.

We have engineered a yeast strain that expresses functional human TRPV1 (hTRPV1), and designed a high-throughput fluorescence assay for ion channel function. The engineered yeast strain has a chromosomally integrated hTRPV1 gene in conjunction with a plasmid expressing GCaMP and MScarlet. In this design, a functionally expressing TRPV1 channel should open in response to agonist application, allowing calcium influx and GCaMP mediated fluorescence. When testing our construct, we observed significant increases in fluorescence upon capsaicin administration. In addition, time course experiments revealed that agonist-induced calcium signaling is dose-dependent and linear, further suggesting that the observed increase in fluorescence due to calcium flux. The functional TRPV1-expressing strain has proven to be of great utility for assaying multiple known TRPV1 ligands, validating its potential for drug screens, and for further dissecting pain pathways.

In conjunction with new methods for machine-learning-based drug discovery that have proven useful for identifying and classifying ligands using yeast-based CB1R and CB2R cannabinoid receptor assays, we believe that it should be possible to develop and refine chemomodulatory strategies for pain that avoid feedback loops leading to addiction.

Genome-Wide CRISPR-Cas9 Screen Reveals a Persistent Null-Hyphal Phenotype That Maintains High Carotenoid Production in *Yarrowia Lipolytica*

Brian Lupish, Jordan Hall, Cory Schwartz, Adithya Ramesh, Clifford Morrison, Ian Wheeldon

Yarrowia lipolytica is a metabolic engineering host of growing industrial interest due to its ability to metabolize diverse energy sources including hydrocarbons, fatty acids, glycerol, and other renewable carbon sources. It also can generate excess acetyl-CoA, which is a precursor to valuable natural products including carotenoids. However, *Y. lipolytica* is a dimorphic yeast which undergoes a transition to a multicellular hyphal state under various stress conditions, such as nutrient or oxygen starvation, as well as pH or temperature shock. The hyphal state can negatively impact biosynthetic activity while increasing culture viscosity, inhibiting oxygen and nutrient mass transfer in liquid culture. Any reduction in nutrient or oxygen diffusion is highly undesirable for industrial bioproduction. Identifying mutations that prevent the formation of hyphae would help alleviate the bioprocess challenges that their formation creates. To this end, we conducted a genome wide CRISPR knockout screen to identify single gene deletions that prevent the transition to hyphal morphology. The screen identified five mutants with a null-hyphal phenotype—*ΔRAS2*, *ΔRHO5*, *ΔSFL1*, *ΔSNF2*, and *ΔPAXIP1*. Of these hits, only *ΔRAS2* suppressed hyphal formation continuously over multiday growth of an engineered lycopene production strain. The others either had reduced growth phenotypes (*ΔSNF2* and *ΔPAXIP1*) or only transient suppression of the hyphal transition (*ΔRHO5* and *ΔSFL1*). Furthermore, the *RAS2* knockout was the only genetic disruption characterized that did not negatively affect lycopene titers over an 8 day production run, producing more than 5 mg L^{-1} OD^{-1} when deleted from a *Y. lipolytica* strain engineered with a heterologous pathway which enhances carbon flux through the mevalonate pathway. These data suggest that a *ΔRAS2* mutant of *Y. lipolytica* could prove useful in engineering metabolic engineering hosts for the production of carotenoids and other biochemicals.

Modeling Recombinant Gene Expression Dynamics across Bacterial Growth Stages

Harikrishnan Namboothiri, Ayush Pandey, Chelsea Hu

Dynamical modeling is a powerful tool for studying synthetic biology as it enables researchers to predict, design and optimize the behavior of synthetic biomolecular networks *in silico*. However, until now, these models have been primarily limited to studying bacterial cells in the exponential (log) growth stage, which requires constant feeding of nutrients for bacterial culture, severely complicating systems design in applications. To overcome this limitation, we developed a chemical reaction network (CRN) model to capture the dynamics of gene expression across all growth stages. We then reduced it to an effective dynamical model with five equations. The effective model approximates the transcription, mRNA degradation, translation, and protein dilution rates as functions of a logistic growth profile, enabling the simulation of gene expression dynamics throughout the entire growth cycle of the bacterial culture. Using this effective model, We have been able to gain insight into previously unexplained gene expression dynamics observed in batch cultures. With this new development, we hope that dynamical models can start being utilized to design and optimize synthetic biological systems with a more accurate representation of growth dynamics, enabling researchers to more effectively predict and optimize system behavior throughout the entire bacterial growth cycle.

Evaluating Synthetic Transmembrane Receptor-Peptide Ligand Interaction in Programed Mammalian Cells

Matthias Recktenwald, Nichole Daringer, Sebastian Vega

Advances in synthetic biology have brought about the ability to program cells with transmembrane receptors which initiate platform activation in response to soluble and cell-laden ligands; however, there are limited platforms capable of activation and negative feedback in response to biocompatible material laden ligands. Such a design would expand the field of cell-based synthetic biology into the field of tissue regeneration. This work utilizes a transmembrane receptor which, in response to an extracellular ligand, phosphorylates a cytoplasmic substrate domain inducing rapid cytosolic protein dimerization which can be reversed using selective phosphatases and monitored using split fluorescent molecules. To-date, the extracellular ligand binding domain has been reengineered to allow for the activation of this platform in response to epitope tag peptides by utilizing *de novo* computationally derived monobody-based receptors. A thiol group is added to these peptides to allow for their conjugation and patterning onto hydrogels using light-mediated thiol-norbornene click chemistry reactions. In both their soluble and hydrogel conjugated forms, these peptides were shown to activate this platform in transiently transfected HEK293 cells as evidenced by a significantly higher mNeonGreen expression in the soluble peptide/receptor matched groups over the no ligand groups and using confocal microscopy. Ongoing work involves expanding the number of ligand binding targets to allow for negative feedback in response to an orthogonal peptide as well as expanding the intracellular domain towards activation of endogenous transcription and secretion of biologically relevant molecules using a split dCas9a platform in place of the split fluorescent domains.

In Situ Monitoring of Cancer Metastasis Dynamics with Acoustic Reporter Genes

Pierina Barturen-Larrea, Mengtong Duan, Margaret Swift, Mikhail Shapiro

Metastasis causes the most deaths in patients with cancer, yet limited tools are available to study the in situ dynamics of the metastatic cascade. We generated mammalian acoustic reporter genes (mARGs) that encode gas-vesicles (GVs), a unique class of genetically encoded air-filled protein nanostructures. GVs are capable of producing nonlinear ultrasound contrast distinguishable from the linear scattering of tissue, enabling genetic labeling of cancer cells for ultrasound imaging with high spatial resolution (<100 μm) and single-cell sensitivity. We hypothesized that cellular ultrasound imaging using mARGs will allow us to track the dynamics of individual metastases in situ. We chose the mouse liver as a model to continuously track individual metastases due to its prevalence as a metastatic site and its accessibility for ultrasound imaging. To generate a liver metastasis model, we injected MDA-MB-231 breast cancer cells expressing mARGs into the mouse portal vein. Single metastases were detected by performing in vivo 3D imaging using nondestructive non-linear pulse sequences alongside B-mode anatomy. We are monitoring the development of mARG-expressing liver metastases during their entire life cycle with ultrasound, and we will use image registration algorithms to facilitate the analysis of single metastases dynamics over time. We are evaluating the accuracy of ultrasound in localizing and sizing metastases by comparing post-hoc ex situ ultrasound and optical images of the liver metastases. We are also engineering other cancer cell lines to express mARGs for spontaneous metastasis studies. Our results suggest that mARGs and ultrasound may make it possible to track in situ single metastases, opening the possibility to use this technology to study the dynamics of metastatic growth to inform new approaches for cancer diagnosis and treatment.

Experimental Demonstration of the Evolution of Operon Formation Mediated by Transposable Element Activity

Yuki Kanai, Saburo Tsuru, Chikara Furusawa

The regulation of gene expression in prokaryotes is characterized by operons, where multiple genes under the same promoter sequence are co-regulated. While the benefit of forming operons have been widely discussed, the mechanism by which operons form thorough evolution remains poorly understood.

Insertion sequences (IS) are DNA transposons ubiquitously found in prokaryotic genomes. Through studying the IS-mediated evolution of *Escherichia coli* using an engineered derivative of a common IS, IS*3*, we observed that IS activity can lead to large deletions of DNA sequences adjacent to the IS. This finding inspired us to propose a new model of operon formation: IS activity facilitates the formation of operons by bringing together two functionally related genes by deleting redundant sequences in between them.

To demonstrate the validity of the proposed model, we developed a fluorescent reporter system to detect the various mutations caused by IS activity and to select cells that formed new operons after evolution by linking a fluorescent protein gene to an antibiotic resistance gene. Surprisingly, we discovered *E. coli* with newly formed operons as expected, after only one day of culture. This is the first study to showcase the evolutionary formation of operons in an easily reproducible manner.

Our proposed model can explain both adaptive and neutral evolution of operons, compatible with our current knowledge about operon formation identified by comparative genomics, and compatible with other models of operon formation. However, the detection of operon formation through the proposed mechanism can be challenging as IS is lost after operon formation. Future studies could analyze ongoing evolution in IS-rich organisms to elucidate the extent to which the formation of existing operons can be attributed to our proposed model.

Exploring Catabolic Electron Flux Design Space for Building Microbial Bioelectronics

Biki Bapi Kundu, Caroline M. Ajo-Franklin

Synthetic Biology enables the building of tunable and controllable microbial biosensors. The microbial biosensors with an electronic signal as output are highly suited for some applications. This electronic signal relies on microbes' ability to reduce a terminal electron acceptor outside the cellular boundaries, known as extracellular electron transfer (EET). In this study, we looked into *E. coli*'s innate ability to carry out EET with the help of externally added redox mediators. Despite the longstanding familiarity with this innate ability, its precise mechanism is yet to be understood. Therefore, we systematically investigated the electron fluxes from carbon catabolism to different electron reserves in the cell. Our investigation provided insights into the potential enzyme and the cell-native redox-active molecule partners of the externally added redox mediators. We found that the deletion of oxidoreductases in the inner membrane and cytoplasm decreased the levels of EET. However, the elimination of membrane-bound dehydrogenases increased the levels of EET. These experiments suggest that electron exchange predominantly occurs in the cytoplasm. Furthermore, we delved deeper into the dynamics of EET with altered NADH and NADPH levels and discovered that the EET levels directly correlate with catabolic-reducing equivalents. Our study establishes a novel *in vivo* approach to screen for redox mediator enzyme partners and provides a set of design rules for electronic biosensing in *E. coli* and other chassis.

Metabolically-Targeted dCas9 Expression in Bacteria

Gregory M. Pellegrino, Tyler Browne, Keerthana Sharath, Khaleda A. Bari, Sarah J. Vancuren, Emma Allen-Vercoe, Gregory B. Gloor, David R. Edgell

The ability to restrict gene expression to a relevant bacterial species in a complex microbiome is an unsolved problem. In the context of the human microbiome, one desirable target metabolic activity are glucuronide-utilization enzymes (GUS) that are implicated in the toxic re-activation of glucuronidated compounds in the human gastrointestinal (GI) tract, including the chemotherapeutic drug irinotecan. Here, we take advantage of the variable distribution of GUS enzymes in bacteria as a means to distinguish between bacteria with GUS activity, and re-purpose the glucuronide-responsive GusR transcription factor as a biosensor to regulate dCas9 expression in response to glucuronide inducers. We fused the *Escherichia coli gusA* regulatory region to the dCas9 gene to create pGreg-dCas9, and showed that dCas9 expression is induced by glucuronides, but not other carbon sources. When conjugated from *E. coli* to Gammaproteobacteria derived from human stool, dCas9 expression from pGreg-dCas9 was restricted to GUS-positive bacteria. dCas9-sgRNAs targeted to *gusA* specifically down-regulated *gus* operon transcription in Gammaproteobacteria, with a resulting ~100-fold decrease in GusA activity. Our data outline a general strategy to re-purpose bacterial transcription factors responsive to exogenous metabolites for precise ligand-dependent expression of genetic tools such as dCas9 in diverse bacterial species.

Designing a Mechanism for Strand Displacement Circuits to Convert Prokaryotic RNA to Heterogenous DNA Sequences

Aaron Stibelman, Melissa K. Takahashi

Synthetic gene circuits are powerful tools used in synthetic biology for applications spanning metabolic engineering to nucleic acid diagnostics. One strategy for their design involves toehold mediated strand displacement (TMSD), to exchange one DNA or RNA strand (input) with another of the same type (output). A caveat to TMSD-based strategies, however, is their inability to produce dissimilar output sequences since their displacement is driven though the hybridization of two fully complementary strands, via Watson-Crick base pairing. This limits the downstream function of TMSD outputs, when they are implemented for the sensing of unchangeable input RNA sequences derived from living cells. As a solution, we have engineered a novel approach to the TMSD mechanism, which can exchange naturally occurring RNA molecules with heterogeneous DNA sequences, used to drive alternative downstream circuit functions. This was achieved by designing a DNA that could rearrange in secondary structure after hybridizing with a specific RNA input. For this application, we have developed a circuit capable of detecting the MicF sRNA, produced in *Escherichia coli*, by converting its sequence into one capable of inducing the transcription of a fluorescent RNA aptamer reporter. These constructs were made with specificity for MicF through the software, NUPACK, which is able to predict nucleic acid secondary structure and interaction. Our initial tests have revealed that the circuit can be induced by low nanomolar concentrations of MicF but experiences some leak due to circuit component misfolding. To improve their functionality, we have incorporated strategies to reduce leak by exploring various modifications to each component's sequence and structure. Although further optimization is required, our preliminary results suggest that this novel TMSD mechanism has the potential for application in circuits requiring the ability to convert specific RNA inputs to alternative DNA outputs.

Yellow Is the New Green: Investigation and Evolution of Old Yellow Enzymes for Sustainable Polymer Production

Jacob Sicheri, David H. Kwan

Biopolymers represent an opportunity to replace petrochemical derived plastics and fabrics with a bio-renewable alternative. Within the class, we're interested in Old Yellow Enzymes (OYEs) due to their broad substrate range and ability to act on many different configurations of alkenes. Due to the broad substrate profile, OYEs are primary candidates for biocatalysts that can generate biopolymers. Using a novel, fluorescence-based, high-throughput assay developed by the Kwan Lab, we are investigating the activity of a phylogenetically and functionally diverse pool of nine OYEs on four biopolymer precursors to identify candidates for enzyme improvement via directed evolution. This improvement is necessary because these OYEs are not efficient enough to overtake traditional polymer synthesis. With a growing global shift towards sustainable technologies, OYE-based biopolymer production is emerging as an opportunity to supplant traditional chemical synthesis.

Current assays to measure OYE activity are flawed. Existing techniques utilize the decrease in NADH/NADPH absorption as OYE acts which has a finite lower limit resulting in inaccuracies. Our assay overcomes these issues by linking OYE activity to fluorescence. Currently, the nine enzyme candidates have been purified have been successfully screened for activity on four biopolymer precursors and top candidates are being investigated further through characterization of enzyme kinetics. Once fully characterized, the most interesting candidate will be improved through directed evolution. Mutant libraries will be generated using multiple potential directed evolution techniques including site saturation mutagenesis. Directed evolution will proceed in a semi-rational manner, taking advantage of known structures where possible. Once mutant libraries are generated, their activities will be assessed via our fluorescence-based assay. This work will contribute to lessening dependence on fossil fuels through the improvement of generating sustainable alternatives.

Design of Synthetic Antisense Oligonucleotides to Modulate sRNA Regulation and Improve Efficacy of Antibiotics in *Escherichia coli*

Min Jen Tsai, Melissa K. Takahashi

Widespread antibiotic resistance in pathogenic bacteria is a rising concern. Reducing membrane permeability is a common mechanism utilized by bacteria to survive unfavorable environments. For example, in the presence of antibiotics, the sRNA MicF is produced and binds *ompF* mRNA, which inhibits outer membrane porin expression in *Escherichia coli*, thus reducing the uptake of antibiotics. To address this, we sought to design antisense oligonucleotides (ASOs) targeting the MicF sequence where it interacts with *ompF* mRNA. Interfering with MicF binding could prevent repression of OmpF and thus increase membrane permeability and the cells' susceptibility to antibiotics. To screen potential ASOs, we developed a cell-free transcription-translation (TX-TL) assay that assesses OmpF expression via a reporter that fuses the 5'-untranslated region of *ompF* with the super-folder green fluorescent protein coding sequence. Peptide nucleic acids (PNAs), a type of ASO, conjugated to cell-penetrating peptides have been used to successfully target RNA molecules in bacteria; however, the length of the PNA is limited to 10-12 nucleotides (nt). The predicted interaction between MicF and *ompF* is 33 nt long, therefore we used the TX-TL assay to screen shorter ASOs for effective MicF sequestration. We were unable to identify a single 10-12 nt ASO that completely sequestered MicF from *ompF* and the best candidate, which targeted the first 12 nt of MicF failed to increase susceptibility to antibiotics in a minimum inhibitory concentration (MIC) assay. Thereafter, we developed a new approach to use two ASOs simultaneously. We specifically targeted the MicF regions predicted to interact with the ribosome binding site and start codon of *ompF*. This dual approach proved successful in both the TX-TL assay and the ability to increase *E. coli*'s susceptibility to antibiotics. Our current findings demonstrate the feasibility of using TX-TL to screen ASO candidates designed to modulate sRNA regulation in bacteria.

Continuous, Label-Free Monitoring of *in Vitro* Cell Culture Progression Using Polymer-Coupled Resonant Sensors

Yee Jher Chan, Samuel Rothstein, Nigel Reuel

Real-time, label-free sensors for *in vitro* monitoring of cell count and quality provides critical information for cellular studies and biomanufacturing applications. To date, real-time monitoring of cell cultures is dominated by microscopy techniques. However, these techniques often require destructive labeling and are not well suited for large-scale manufacturing applications. Here we introduce an alternative technique that uses an acrylic polymer that softens in response to secreted molecules during cell growth. The mechanical change is transduced into an electrical signal through a wireless resonant sensor, thereby maintaining system sterility. When culturing HeLa cells in a culture dish containing 7mL of media, the sensor reports over 30 MHz shift in resonant frequency over four days of cell growth (signal to noise ratio of >100) which was correlated to cell count via microscopy. In addition, treatment with drugs that induce cell death resulted in an instantaneous halt to the resonant frequency shift. The sensor response was also observed with Jurkat, K562, T cells, Chinese hamster ovary cells, and *E. coli*, suggesting that the sensor is applicable for prokaryotes and eukaryotes as well as both adherent and suspension cells. The sensor has been successfully demonstrated in petri-dishes, well-plates, and single use bioreactors as method to track cell growth. On-going work to determine identity of the transducing secreted molecules will be discussed as well as selective targeting of other biomolecules using this responsive polymer based transduction technique.

Characterization of Fluorescent Aptamers As Tools for Dynamic Tracking of Native RNAs in Bacteria.

Alice Grob, Annie Tang, Elisa Franco, Francesca Ceroni

Tracking RNA production in live cells is key to understand the dynamics of many cellular regulation pathways and strengthen our ability to exploit these for the development of synthetic biology control tools. One example is bacteria cellular burden response to heterologous expression. Understanding burden dynamic regulation would be essential to further improve tools mitigating it. Fluorescent RNA aptamers have proven to be useful tools for tracking RNA production. However, they can only be adopted downstream of strong viral promoters (i.e. T7, T5) as their signal intensity is too low under natural promoters. This greatly limits our ability to easily investigate the dynamics of native RNA expression. Recently, stronger aptamers were published, among these two novel broccoli aptamers and the pepper aptamer. These have so far been adopted in mammalian cells and in vitro systems, but not in bacteria. Here, we start by characterising these aptamers in E. coli and benchmark their features against the standard broccoli aptamer already adopted in bacteria. We characterise their fluorescence signal when they are placed under different promoters, comparing expression from T7 and native promoters (e.g. pBad, pLux) to investigate if these aptamers can allow in vivo RNA tracking in bacteria with greater precision. First, we compare expression at a define time point. Then, we show their behaviour over a 5-hour window, by plate reader and flow cytometry assays, to provide information on the best protocol to adopt. We will continue by comparing expression in bacteria to expression in mammalian cells and in vitro, to provide a comprehensive understanding of the features and limitations of these aptamers. Finally, these aptamers will be adopted to develop biosensors for burden and native stress-regulated small RNAs. Our characterisation framework will prove useful to advance measurements and tool development in synthetic biology where RNA tracking is sought.

A Bioinformatic Pipeline Supporting the Selection of Bacterial Promoters with Target Features from the Analysis of Transcriptomic Data

Debora Dallera, Stefano Quaranta, Davide De Marchi, Giovanni Cipriano, Paolo Magni, Lorenzo Pasotti

The increasingly demand of expanding the libraries of regulatory parts for non-model microorganisms motivated the development of novel strategies based on bioinformatics approaches for their *in-silico* prediction as an alternative to the high cost and time-consuming *in-vivo* search and characterization [1]. A new bioinformatics pipeline has been developed with the aim to detect constitutive endogenous promoters in microorganisms with annotated genomes, based on transcriptomics data in different experimental conditions. The pipeline is divided into two procedures. The first procedure analyzes transcriptomic data from microarray and RNA sequencing (RNA-seq) experiments. Data are downloaded from the NCBI Gene Expression Omnibus (GEO) database and then processed to obtain a list of genes with stable transcription level, i.e., non-differentially expressed (NDE) genes, across the conditions. Using expression values from both technologies increases the reliability of the result about NDE genes as many more experimental conditions are considered from publicly available data repositories. The second procedure predicts Transcription Start Site (TSS) upstream of NDE genes from RNA-seq data. Thanks to the RNA-seq resolution, it is possible to reconstruct reads coverage profiles base by base along the genome inferring the TSS genomic coordinate. Genome-scale identification of promoters with stable activity in *Escherichia coli* is provided as pipeline application. The validation of the first procedure is done by selecting a subset of genes with different expression levels and verifying that, in the literature, they are annotated as NDE genes in various conditions. The performance index of the second procedure is based on TSS comparison with promoter annotated in EcoCyc database. The pipeline can allow to expand the toolkit of biological parts implemented in bacterial synthetic circuits to enable reliable engineering and gene expression control in model and non-model species.

1. Gilman and J. Love, "Synthetic promoter design for new microbial chassis", *Biochemical Society Transactions*, 44(3), pp. 731-737, 2016.

Integration of Post-Translational Circuit with Native Signaling Pathways

Evan Hutt

Cell-based therapies can overcome the limitations of current cancer treatments that attack the cell cycle with little specificity, resulting in severe off-target effects. Advancements in chimeric antigen receptor (CAR) T-cells and similar systems improve the safety and efficacy of targeted cancer treatments, but still rely on intracellular native signaling mechanisms and offer little downstream control. These caveats allow for potential crosstalk with other signaling pathways and T-cell exhaustion. Optimized integration of native signaling pathways with a synthetic construct has the potential to insulate cellular response from crosstalk with signaling pathways activated by the tumor microenvironment, enabling direct control over the cellular response and inhibition of T-cell exhaustion. Modular synthetic constructs have been used to release a synthetic transcription factor in response to interchangeable scFv-ligand binding through protease mediated cleavage, but post-cleavage, this system is irreversible[1]. Conversely, post-translational circuits utilize kinase-phosphatase mediated protein-protein interactions to dictate other protein-protein interactions. Preliminary feasibility and characterization of this platform has been performed by inducing the interaction of split mNG via rapamycin binding receptors transfected in HEK293 cells. Split fluorescent proteins and scFv's will be replaced with proteins of native signaling pathways to establish a specific point of integration between the post-translational circuit and the native pathway, in response to a novel ligand. The post translational circuit described in this study will be integrated with key points of native pathways important in the immune response to cancer such as NF-κB and Stat6 to provide desired control over cell function. Moreover, this platform establishes an avenue to create new cell-based therapies for other conditions mediated by known signaling pathways, increasing the applicability of the approach.

1. Daringer et al. ACS Syn Bio. 2014

Modular Assembly of Fusion Domains for High Throughput Polymerase Engineering

Inyup Paik, Brice Graham Hendrickson, Jonghyeok Shin, Hannah Do, Sanchita Bhadra, Andrew D. Ellington

The properties of proteins can be altered through appending fusion domains. While typically these domains have been used to increase solubility or stability, the fusion of nucleic acid binding domains to polymerases is especially interesting as it has been shown to improve processivity (e.g., Phusion DNA polymerase). Despite its longstanding use in protein engineering, there is currently no standardized cloning and rapid testing platform for introducing fusion domains, including to polymerases. Herein, we have developed a Golden Gate assembly-based fusion domain engineering strategy, along with a rapid protein activity screening platform that uses 96-well-based, preparation-free assays. Using this platform, we have evaluated 15 reverse transcriptases (RTs) in combination with four distinct linkers and 24 nucleic acid-binding fusion domains from various sources. After assembling and testing approximately 500 unique fusion RTs, we identified synthetic fusion RTs that exhibited thermostability as well as improved processivity in various nucleic acid-related assays and diagnostics. In parallel efforts, we have generated fusion domain libraries of T7 RNA polymerase, and begun to again demonstrate the production of enzymes with improved processivity during transcription. These tools should prove of great utility for generating enzymes that have higher performance in diagnostic assays, DNA assembly, and ultimately in cell-based synthetic biology applications.

Leveraging Pathogen Vulnerabilities to Increase Antibiotic Efficacy and Limit Infection Relapse

Travis LaGree

Antibiotic tolerance continues to be a major driver for antibiotic treatment failure and the development of antibiotic resistant pathogens. To target antibiotic-tolerant bacterial populations more effectively, understanding the mechanisms that underlie their resilience is crucial. Nutrient-depleted environments, which often mirror physiological niches where pathogens thrive, can slow bacterial growth and foster the emergence of antibiotic-tolerant bacteria. We recently demonstrated that metabolic reprogramming through the supplementation of carbon sources can sensitize slow-growing uropathogenic *Escherichia coli* to topoisomerase inhibitors, such as Levofloxacin, Delafloxacin and Zoliflodacin. Crucially, we found that metabolically stimulated *E. coli* restart transcription and translation, which increases active gyrases that these antibiotics can target. Furthermore, in *E. coli* that can tolerate these compounds, DNA damage response genes, such as the SOS response regulator, *recA*, are needed to effectively recover and resume out-growth after antibiotic treatment terminates. These findings have potential downstream applications in the pursuit of sensitizing human pathogens to existing antibiotics. Specifically, we look to modulate the microbiome within infection niches to suppress pathogens. Our understanding of tolerant bacteria's metabolism guides the design of potential strategies for modulating the microbiome and sensitizing the pathogens to antibiotics. Additionally, utilizing our understanding of antibiotic sensitivity and tolerance, we can construct commensal/probiotic strains that are tolerant to topoisomerase inhibitors. Over time, these engineered strains may more effectively recover, resume growth, and out-compete human pathogens following antibiotic treatment. Successful development of these approaches will enable us to manipulate infection environments to sensitize harmful bacteria to antibiotics and prevent recurrent infections.

Developing *Pichia Kudriavzevii* Strain YB-431 As a Chassis for Energetically Competitive Production of Biochemicals

Isabella M Spielmann, Andrei S Steindorff, Joonhoon Kim, Bruno Klein, Pahola Thathiana Benavides, Ryan Davis, Igor Grigoriev, Jon K. Magnuson, Kyle Pomraning, John M. Gladden, Di Liu, Nathan J Hillson, Henrique De Paoli

Pichia kudriavzevii is a non-conventional and usually diploid yeast, also known as *Issatchenkia orientalis* (teleomorph) and *Candida krusei* (anamorph), which shows unique traits valuable in developing solutions for energy and materials sciences. One such biotechnological application is the production of organic acids (e.g., succinic acid, itaconic acid), which can be accomplished under low-pH conditions (e.g., pH 2.0), however strain-specific traits can make isolates distinct from each other and unique to certain applications (e.g., strain MBY1358, which grows at 44°C). Here we present the diploid genome sequence of YB-431 (https://mycocosm.jgi.doe.gov/Pickudri1/Pickudri1.home.html) along with aerophilic and microaerophilic growth data, using different carbon sources (including pre-treated plant biomass) and temperature (30°C and 42°C) to assess variables relevant for biorefinery processes. YB-431 biomass was comparable between micro- and aero-philic conditions when grown at 30°C despite significant differences in the percentage of dissolved oxygen. Also, up to 74% of YB-431 maximum growth at 30°C was maintained at 42°C. Both data combined suggests YB-431 dispenses oxygen compressor and extended broth chilling in large scale fermentation making it more suitable to environmentally and energetically competitive fermentation due to cost reduction. We also present a set of 15 different promoters isolated from YB-431 and their heterologous expression of CFP under different conditions in both *Pichia kudriavzevii* and *Saccharomyces cerevisiae*, which can advance metabolic engineering strategies requiring minimal endogenous regulation of organisms' native promoters.

Rapidly Assemble Genes in Your Laboratory Using Automated Enzymatic Oligo Synthesis

Steven Quistad, Anne Céline Derrien, Florence Mahé, Xavier Godron

Gene assembly is a central application to a wide range of scientific fields from vaccine development to synthetic biology. Most gene assembly methods utilize chemically synthesized gene fragments as a starting point which must be ordered from a commercial vendor. This legacy service is prone to long delays when sequences are complex or when oligo supply chains are disrupted, resulting in uncertain research timelines.

The SYNTAX system is an automated oligo synthesis platform powered by Enzymatic DNA Synthesis (EDS) technology that can print oligos directly in your lab within a matter of hours. This all-in-one benchtop solution enables researchers to design and assemble 1-2 kb genes in three days compared to waiting weeks using current commercial oligo suppliers.

To demonstrate the utility of the SYNTAX system in gene assembly. The 1.7 kb Influenza A hemagglutinin gene was used as a model system. Three double stranded DNA (dsDNA) blocks were generated from EDS-synthesized ssDNA oligos using the PCA approach followed by error correction. The dsDNA blocks were then assembled using a commercially available kit, transformed into BL21 competent cells and colonies were selected for Sanger Sequencing confirmation.

By providing the user with complete control over their oligo synthesis pipeline the SYNTAX system can ensure consistent productivity and accelerate the iterative design-build-test cycle.

Engineering a Metabolic Pathway for Naringenin Biosynthesis in a Xylose-Fermenting *Saccharomyces Cerevisiae*

Sun-Mi Lee

Naringenin is a widely used flavonoid in the food, pharmaceutical, and biomedical industries. Microbial cell factories offer an economical and environmentally friendly production platform for naringenin, in which using lignocellulosic biomass as feedstock further improves sustainability. Here, *Saccharomyces cerevisiae*, one of the most preferred industrial hosts with GRAS(generally regarded as safe) status, was engineered to become a platform host for naringenin production using a renewable feedstock of biomass. Specifically, a heterologous naringenin synthetic pathway was introduced into an engineered glucose/xylose co-fermenting *S. cerevisiae*. The naringenin production was then further improved by adapting *de novo* pathways with the genes sourced from various organisms. The optimization of the naringenin biosynthesis pathway resulted in a seven-fold increased naringenin titer. Our results highlight the importance of the synergetic combination of heterologous enzymes in a *de novo* pathway and the potential of glucose/xylose co-fermenting strain for naringenin production.

Enhanced Rolling Circle Amplification Performance with a Newly Engineered phi29 DNA Polymerase

Kyle Vrtis

Isothermal DNA amplification techniques, such as Rolling Circle Amplification (RCA), exponentially increase the amount of DNA in a sample without the constraint of thermocycling. RCA has become an invaluable tool for many biotechnology applications including DNA sequencing, cell-free DNA synthesis, and cell-free protein expression. The most widely used DNA polymerase for RCA is phi29, which possesses high processivity and fidelity, and strong strand-displacement activity. We have engineered a new phi29 DNA polymerase, phi29-XT, that shares the positive characteristics of the wild-type phi29, but also generates higher yield with greater sensitivity than the wild-type enzyme. Additionally, while wild-type phi29 works optimally at 30°C, phi29-XT performs best at 42°C. Herein, we demonstrate the utility of phi29-XT DNA polymerase-mediated RCA for applications such as cell-free protein expression and high-throughput fosmid amplification directly from bacterial cells. Given the improved performance of phi29-XT over the wild-type enzyme, phi29-XT should be considered as a replacement DNA polymerase for many RCA workflows.

Engineering Recombinase-Based Synthetic Memory

Dowan Kim, Andrew Short, Brian Huang, Corey Wilson

Engineered genetic circuits have allowed to develop myriad applications – e.g., biomanufacturing, biosensors, and living therapeutics. An increasingly important genetic circuit technology comes in the form of synthetic genetic memory engineered for living systems, used to record transient events such as exposure to chemicals or changes in environmental conditions. Early synthetic genetic memory utilized recombinases that recombine specific DNA elements resulting in permanent change – i.e., the inversion or excision of a cognate DNA sequence. One of the limitations of existing recombinase-based memory is the small number of recombinases that can be regulated distinctively. In addition, to fully deploy synthetic recombinase-base memory several challenges need to be addressed – e.g., validating orthogonality between recombinase functions, intercellular noise that interferes with desired functions, and unintended effects related to recombinase recognition to attachment sites. The ability to engineer synthetic memory programs systematically is predicated on the reliable regulation of recombinase functions. We posit that the use of transcription factors can facilitate perfect control. Accordingly, we have implemented recombinase-based synthetic memory with the aforesaid properties. Namely, we have developed a programmable synthetic memory array for the general regulation of six recombinases in a single chassis cell. In this system, recombinase activities are controlled by six orthogonal transcription factors that can process disparate inputs – i.e., forming the corresponding synthetic memory array. Leveraging this, generalizable memory platform can be achieved enabling the rapid design of myriad memory programs. In addition, we have introduced a novel strategy and design that enables an end-user to block a given recombinase's function post-translation. Here we repurpose unit operations from our transcriptional programming to achieve a marked increase in the rate of DNA inversion and deletion. Together our synthetic memory technologies for living systems represents a paradigm shift in synthetic biology and will enable a host of new applications.

Pool of Long Amplified Reads (POLAR) Sequencing Reveals Genotypes of Rearranged Synthetic Clusters in *S. Cerevisiae*

Klaudia Ciurkot, Anastasiya Malyshava, Xinyu Lu, Tom Ellis

Synthetic gene clusters are created by the relocation of pathway genes from their original loci to the same locus either with native or heterologous regulatory elements. The cluster can be rearranged and modified by lox-P (or lox-Psym) mediated evolution (SCRaMbLE) which allows for deletion, duplication, translocation, or inversion of the cluster's genes. These new genotypes can exhibit desired features such as improved growth or production. However, the determination of genotype after SCRaMbLE remains challenging due to the scale of possible rearrangements and lack of prior knowledge about them. To tackle this limitation, we propose Pool of Long Amplified Reads (POLAR) sequencing. POLAR sequencing relies on ultra-long (20-30 kb) PCR amplification of a synthetic cluster followed by nanopore sequencing. Long-read sequencing allows to capture reads which cover the whole cluster and to fully determine the genotype of the rearranged cluster. To increase the throughput of discovered genotypes, we combined POLAR sequencing with FACS sorting of cells that underwent SCRaMbLE. By applying this approach to study minimal cell cycle cluster, we were able to discover over 130 unique genotypes. In addition, we used POLAR sequencing to determine a modified histidine cluster which restored cellular growth.

Alternative Carbon Sources in Bioproduction Using Dynamic Metabolic Control of *E. coli*

Taylor Lundy, Zhixia Ye

Metabolic engineering of microbes to become biosynthetic production factories offer an appealing sustainable alternative to chemical manufacturing of keystone molecules used across several industries. Specifically, fermentation of Escherichia coli (*E. Coli*) is the most promising due to its malleable genome and low cost to maintain. In recent years, fermentation of microbes has made tremendous advancements that have made it a realistic option for scalable and sustainable green chemistry. Our platform at DMC Biotechnologies allows for a cost-effective way to produce the chemicals the world needs in a manner that is environmentally friendly. We have developed a standardized two-stage bioproduction platform using dynamic metabolic control, which uncouples cell growth from bioproduction using dynamically controlled gene silencing and targeted proteolysis. By limiting the ability of the microbes to respond to their environment, we improve the robustness and scalability of production, regardless of the bioproduct. To further expand and complement our DMC technology platform, we are exploring the use of different feedstocks as carbon sources. Traditionally, glucose is the main carbon source for *E. Coli* and used heavily in fermentation. However, this means that the glucose used for fermentation is also pulling from our food supply. We are working to expand the carbon source use to include an alternative carbon source that can be converted to biomass at industrial levels. To achieve this goal, ethanol is an appealing substitute for glucose. We have demonstrated a proof-of-principle concept strain using our engineered and dynamically controlled *E. coli* that is able to grow using ethanol as the sole carbon source and convert that ethanol to our products. This is a very promising platform to further reduce the impact that bioproduction has on the world and demonstrates that alternative carbon sources are a viable alternative to traditional fermentation.

Optimized CRISPR-Cas9 Genome-Wide Screening in *Yarrowia Lipolytica* enables the Identification of Genes Essential to Acetate Metabolism

Nicholas Robertson, Varun Trivedi, Adithya Ramesh, Marcus Harland-Dunaway, Sangcheon Lee, Ian Wheeldon

Yarrowia lipolytica is a promising industrial host as it can produce high quantities of lipids, has a large acetyl-CoA pool available for the synthesis of desired metabolites, and can consume a range of low-cost waste substrates, including acetate. Acetate is an important carbon source because it does not compete with agricultural production of commodities as is the case with sugars derived from corn and other crops. Recent advances in electrocatalysis have resulted in carbon capture methods that can use electricity to produce acetate at high efficiencies and purities. Here, we design and deploy an optimized genome-wide functional genomic screen to enhance our understanding of acetate metabolism in *Yarrowia*, with the ultimate goal of identifying genetic mutations that will enhance acetate metabolism and the conversion of acetate into desired oleochemicals. An ideal CRISPR-Cas9 guide RNA library covers all protein-coding sequences with a small set of highly active guides. We report here on our work in optimizing a previously developed CRISPR-Cas9 knockout library, reducing its overall size while increasing the activity of each guide in the library. We accomplished this by generating genome-wide profiles of guide activity and subsequently using this information to remove low activity guides, replacing them with guides predicted to have high activity. To validate our screen, we grew our knockout library on acetate-rich media to discover non-obvious genes involved in acetate tolerance and consumption. Twenty-five genes were validated for their predicted fitness in acetate, fifteen of which had significant change in growth on acetate media. This work demonstrates the ability of our CRISPR-Cas9 genome-wide screen to rapidly discover genes for metabolic engineering.

Engineering Peptides for the Disruption of an *Escherichia coli* Small RNA Regulator Involved in Antibiotic Resistance

Edwin Ortiz, Melissa K. Takahashi

In *Escherichia coli*, outer membrane porin F (OmpF) is used for nutrient uptake, but certain antibiotics can also use OmpF to enter the cell. OmpF expression is regulated by the small RNA MicF. When the cell senses antibiotic invasion, MicF is activated, binds to *ompF* mRNA, and inhibits OmpF expression, making the bacterial cell resistant to antibiotics. Therefore, the aim of this work is to generate a peptide with the highest binding affinity for MicF possible, preventing it from complexing with *ompF* mRNA. To generate the peptide candidates, the arginine rich motif (ARM) of the N protein from bacteriophage lambda was used. Specific regions of the N protein were mutated by randomization, using PCR, to produce peptides with an increased affinity for MicF. Peptides were chosen from a pool of candidates based on fluorescence and absorbance measurements taken from an assay that uses GFP production as a proxy for OmpF expression. Three peptide candidates were identified as the top MicF inhibitors. Molecular dynamics simulations of MicF complexed with each candidate were performed, followed by binding free energy and protein-RNA contact analyses, to design second generation peptides with a potentially higher binding affinity for MicF. A minimum inhibitory concentration assay of the top candidate was performed to determine its efficacy in reducing the concentration of an antibiotic needed to inhibit the growth of *E. coli*. A successful candidate would provide a powerful tool in combating antibiotic resistance. Additionally, given the structural similarity between regulatory RNAs, this approach can be modified and applied to disrupt regulation in other sRNA/mRNA pairs.

Developing Optogenetic Tools in *Shewanella Oneidensis* MR-1 for Conductive Biofilm Patterning and Extracellular Electron Transfer Regulating

Fengjie Zhao, Marko S. Chavez, Kyle L. Naughton, Christina M. Niman, Joshua T. Atkinson, Jeffrey A. Gralnick, Mohamed Y. El-Naggar, James Q. Boedicker

Electroactive bacterial biofilms can function as living biomaterials that merge the functionality of living cells with electronic components. However, the development of such advanced living electronics has been challenged by the inability to control the geometry and uniformity of electroactive biofilms relative to solid-state electrodes. Optogenetics has been developed for spatiotemporal control of gene expression in bacteria in response to illumination with light. Here, we developed a lithographic strategy to pattern conductive biofilms of *Shewanella oneidensis* by controlling expression of the aggregation protein CdrAB with the blue light-induced genetic circuit pDawn. This controlled deposition enabled *S. oneidensis* biofilm patterning on transparent electrode surfaces and electrochemical measurements allowed us to demonstrate tunable conduction dependent on pattern geometry and quantify the intrinsic conductivity of living biofilms.

To further expand the optogenetic tools in *Shewanella*, we developed a red light-induced genetic circuit in *S. oneidensis* based on a reported iLight system. Together these two new genetic tools enabled new experiments to deepen our understanding of current generation and extracellular electron transfer within *Shewanella*.

Overall, we introduced the optogenetics in *Shewanella* and demonstrate the implications of these tools to study extracellular electron transfer and develop new bioelectronics.

Discovery of a Novel Potential Biological Redox-Switch Cytochrome

Sol Choi, Daniel R. Bond

Microbes can alter gene expression in response to specific stimuli, such as light, temperature, or changes in the concentration of soluble compounds. For example, systems are known that sense redox-active molecules including O_2, H_2O_2, NO, CO, sulfides, NADH, and membrane quinones. However, no biological mechanisms are known that can directly sense environmental redox potential independent of a specific proxy molecule. New evidence indicates that a novel family of extracellular cytochrome-based sensor domains activate gene expression specifically in response to redox potential. A model for this family is the BccS/BccR two-component system in the metal-reducing bacterium *Geobacter sulfurreducens*. RNAseq, mutant analysis, and promoter fusion data shows that BccS, a cytoplasmic membrane histidine kinase linked to a periplasmic diheme *c*-type cytochrome PAS domain, is the sole input for the BccR regulator which regulates expression of cytochromes essential for growth when redox potential is low. For example, using a NanoLuc luciferase reporter, induction dependent on both BccS and BccR occurs when redox potential drops below about -0.15 V vs. SHE. Induction is not dependent on the concentration of redox-active molecules such as Fe(II) or cysteine in the medium, but only the ratio between the total oxidized and reduced species. Reductants with higher redox potentials, such as Mn(II) or ascorbic acid, do not trigger induction. While many homologs of the BccS sensing domain are found in linked to methyl-accepting chemotaxis cascades and other complex regulatory networks, both proteins in the BccS/BccR system have simple architectures amenable to modular domain swapping. This discovery enables design of synthetic redox biomonitors able to integrate multiple environmental signals into a single response that will be useful in many sustainable bioelectronic and environmental applications.

Capturing Rare Earth Elements with Bioengineered Spores

Julia Puffal, Lizette Cardenas, David Walker, Inyup Paik, Edward Perkins

Rare earth elements (REE) are essential components of advanced technologies such as hard drive magnets and laser components. However, current approaches for recovering and purifying REE are energy and solvent-intensive, and have limited ability to separate different REE from each other. To address this issue, Lanmodulin (LanM), a protein with a high affinity for REE, has been proposed as an effective tool for REE sequestration and purification. To sequester REE, multiple display platforms can be used to hold LanM proteins, such as spores, which offer a promising platform for tailored extraction of critical metals. *Bacillus subtilis* spores, in particular, are a natural platform that can be genetically manipulated for displaying proteins for various applications, including drug delivery and environmental uses. In this study, we expressed LanM on the surface of *B. subtilis* spores to determine their efficacy in sequestering REE. However, the spore's coat itself can also bind REE, potentially interfering with the selective binding of surface-expressed proteins. Therefore, we explored the ability of both wild-type and LanM-expressing spores to bind dysprosium (Dy^{+3}) and identified the optimal conditions for capturing lanthanides in a solution. We evaluated the effects of pH and salt concentrations on both spore types and established ideal buffer conditions for adsorbing and detecting Dy using a novel multiplex luminescent assay. Overall, our findings suggest that LanM-expressing spores have a high affinity for REE and can potentially be used for the efficient recovery and purification of these critical metals.

Microfluidic Encapsulation of Biological Materials Enables Rapid Kinetics and Screening Approaches

Steven Hoang-Phou, Chao Liu, Congwang Ye, Mariam Mohagheghi, Nicholas N. Watkins, Nikki Hammond, Emma Laurence, Ted A. Laurence, Matthew A. Coleman

Biologically relevant systems are often multi-factorial and demands ways of addressing their inherent complexity. Dimension reduction or simplification of model systems is commonly used to reduce complexity due to resource and cost concerns, however, this risks losing meaningful data. High-throughput microfluidics screens have the potential to address these highly complex systems without sacrificing resolution. Microfluidics-based systems for encapsulation allows us to generate $>10^4$-10^5 capsules of tunable diameters, shell thicknesses, and potentially pore sizes. Here, we combined cell-free protein synthesis (CFPS), microfluidic encapsulation, and fluorescence correlation spectroscopy (FCS) to 1) produce model fluorescent proteins and protein complexes within microcapsules, and 2) measure their diffusion characteristics as well as binding kinetics in real-time. These assessments provided information on protein concentration, protein complex size, and how they interacted. Furthermore, we incorporate a biocompatible and UV-crosslinkable polymer shell that enables long-lasting, stable, storage of the capsules at temperatures ranging from 0° to 37°C. Using this hard polymer formulation, we have also encapsulated the fresh-water alga, *Chlamydomonas reinhardtii*, and observed its persistence for multiple days. We envision that by tuning the capsules' physical properties, it could be possible to support long-term growth of encapsulated cells inside. Further system optimization would make it highly scalable and amenable to $>10^5$ diversity libraries/experimental conditions and will facilitate more efficient screening approaches.

Engineering Mutation Rates of an Orthogonal DNA Polymerase to Enable Deep Exploration of Functional Protein Sequence Space

Gordon Rix

The ability to generate large datasets of diverse, functional protein sequences *de novo* may help identify new starting points for engineering efforts, generate new insights into the protein sequence-function relationship, and inform machine learning for protein design. Continuous evolution, in which user-defined DNA is autonomously mutagenized and replicated in living cells, provides the scale and depth in evolutionary search necessary to achieve such datasets. To this end, we engineered an orthogonal DNA replication system in yeast to durably hypermutate its target plasmid at a rate exceeding 1-million-fold higher than the host genome replication machinery. We then used this system to evolve an enzyme at enormous scale. By serially passaging many small, parallel yeast populations encoding the tryptophan synthase β-subunit (TrpB), we evolved hundreds of thousands of unique, functional orthologs of this enzyme in 3 months. These orthologs are separated by an average pairwise hamming distance of ~35 amino acids, including over 4000 sequences pairs differing by 60 amino acids (15% divergence). This level of divergence matches that between orthologous mouse and human genes, which took ~90 million years of natural evolution to reach.

Prediction of Mixing Time By Rotational Speed of Bioreactor

Hyunwoo Lee, Youngjin Seo

Bioreactor is a device that produces a desired product by growing cells under an appropriate environment after introducing a medium. Methods of culturing cells using the bioreactor is used in various industries, and related research is in progress. In this study, the bioreactor was operated with orbital rotation, and the rotational speed was controlled. For reliability of numerical analysis, results of experiment and numerical analysis were compared. Mixing time was predicted by varying the rotation speed and scale using the verified numerical analysis model.

Building an Orthorep Toolkit

Ria Deshpande, Chang Liu

OrthoRep is a platform in yeast for continuous directed evolution. It utilizes a pair of cytoplasmically located linear plasmid that are orthogonal to the yeast nuclear genome. The gene of interest to be evolved can be transformed onto p1, and an error-prone polymerase responsible for replicating p1 is located in the genome. This polymerase makes on average a million times more mutations than the native yeast polymerase. However, the orthogonality of the polymerase is such that it only replicates p1 and does not interact with the nuclear genome at all. The other plasmid, p2, is responsible for maintenance of the p1 plasmid. As OrthoRep becomes more popular as a way to perform directed evolution at a faster rate, we have a developed a toolkit. The toolkit consists of a set of most used parts and a detailed set of protocols that will enable any lab to get started with using OrthoRep.

Bench-Scale Yield Optimization of Ricinoleic Acid from a Genetically Engineered Strain of Oleaginous Yeast

Jessica Paradysz, Nathan McDonald

Ricinoleic acid (RA) is a fatty acid with high importance within the oleo (oil)-chemical industry. This compound has the ability to undergo a wide range of chemical reactions resulting in numerous derivatives, making it an essential precursor for compounds of interest, such as polyamide polymers, nylons, lubricants, and fuels. Ricinoleic acid is primarily extracted from the castor bean of the castor plant, traditionally grown in tropical regions of the world. However, this process comes with climate and agricultural impacts as well as high risks to workers due to the accumulation of ricin toxin within castor beans. Recent efforts have demonstrated that ricinoleic acid can be produced microbially in genetically engineered strains of oil-producing (oleaginous) yeast. While yeast can produce this critical molecule, titers that are relevant to industrial biomanufacturing applications have not yet been achieved. To date, RA production in oleaginous yeast has been performed through genetic engineering of the strain *Yarrowia lipolytica*, through introduction of the transgene CpFAH12 hydroxylase. Through genetic modifications, hydroxylation of oleic acid resulting in RA production has been productive up to 60mg of RA per gram cell dried weight. Through a Design of Experiments (DOE) statistical approach, media components, such as carbon and nitrogen sources, pH, and temperature will be strategically tested in order to determine the optimal growth conditions for the genetically engineered strain of *Y. lipolytica.* Bottlenecks identified throughout media optimization will be mitigated through additional metabolic engineering to optimize the RA titer output. Through developing a feedback loop between genetic modifications and fermentation conditions, paired with RA purification strategies, the biomanufacturing of this key molecule can outcompete current methods.

High Throughput Test Tools and Applications in Industrially Relevant Micro-Organisms Such As *P. Putida* and *C. Glutamicum*

Herbert Huttanus, Jeanette Velasquez Guzman, Ramesh Kumar Jha

The major breakthroughs in synthetic biology are typically pioneered in well-characterized model organisms and are touted for their potential in industrial biomanufacturing applications. However, these model organisms are not always well adapted to the conditions required for large scale fermentation. Luckily, there are virtually unlimited "non-model microorganism" that can serve as industrial platforms that are more robust or have unique and versatile metabolisms that make them attractive alternatives. To fully utilize these non-model organisms, however, they must be "domesticated" by adapting synthetic biology tools for their intended host. In this work, we explore the adaptation of metabolite biosensors to two industrially relevant non-model organisms, *Pseudomonas putida* and *Corynebacterium glutamicum*. In *P. putida,* a biosensor was employed to screen a mutation library and ultimately, enhance the production of cis,cis-muconic acid, a precursor compound to various industrial polymers including nylon. The mutation library consisted of semi-rational alterations to the gene *hexR*, a global regulator known to influence muconate titers. Mutations included substitutions to the inducer binding domain and DNA binding domains of the *hexR* transcription factor so as to create new regulatory combination profiles with the potential for more optimized muconate production. *C. glutamicum* is another industrially relevant microorganism that has long been used for the production of amino acids. Despite its industrial impact, there is still a need for more "domestication" to expand its capabilities. The biosensor for muconate in *P. putida* was adapted for use in *C. glutamicum* and an additional biosensor for chorismic acid was also developed. Chorismate is a highly branched metabolic node that leads to several industrial biosynthetic pathways, including muconate. Both biosensors were tested for use individually, and as a pair (dual sensing). Dual sensing with biosensors expressing different fluorescent proteins can facilitate a deeper understanding of (and screening for) relative carbon flux through linked pathways.

Membrane Proteins As Privileged Scaffolds for Gene Entanglements

Christina Kang-Yun, Tiffany Halvorsen, Sean Leonard, Jose Manuel Martí, Yongqin Jiao, Dan Park

Engineered bacteria with synthetic biology designs can rapidly lose their desired functions due to genetic instability when deployed into natural environments. Moreover, these synthetic designs can spread into the native bacterial population via horizontal gene transfer. Thus, drastic improvement in genetic stability and biocontainment strategies must precede environmental application of these engineered bacteria. Here, our goal is to establish a generalizable method for gene entanglement – encoding two genes within different coding frames – to improve stability and prevent horizontal gene transfer of synthetic designs. Genes with greater sequence flexibility are more amenable to gene entanglement, though protection against mutations of such designs could be subpar. We posit that membrane proteins (MPs) are attractive entanglement partners, as entanglement of a gene of interest (GOI) into an essential MP may provide a balance between flexibility and mutational robustness. Here, we used CAMEOX – a computational biology pipeline for generating gene entanglements – to assess the entanglement feasibilities of MPs. 32 α-helical inner MPs and 22 β-barrel outer MPs in *E. coli* MG1655 were tested as entanglement partners with four type II toxins (*hicA*, *relE*, *ccdB*, *parE*) and two conditionally essential genes (*panD*, *purE*). Based on the fitness scores determined by CAMEOX, *hicA* was selected as the GOI and *bamA*, *lptD*, *cmlA*, and *tetA* were selected as MPs to experimentally screen for fitness. Two functional *hicA* variants were identified from 77 designs in a preliminary screen, and their sequence identity to wild-type HicA was as low as 60.3%. A large-scale screening of *hicA* variants will be performed, and the corresponding MP partners of resulting functional *hicA* variants will be screened to identify successful entanglement designs. The results of this study will determine the feasibility of using MPs as a scaffold and improve our understanding of the key design considerations in gene entanglements.

Design and Synthesis of a Yeast Transcription Factor Module

Daniel Lusk, Zachary Krieger, Alessandro Luis Venega Coradini, Yiwei He, Chenghao Du, Cara Hull, Oscar M. Aparicio, Ian M. Ehrenreich

Chromosome synthesis can be used to reorganize genomes in ways that help improve understanding of cellular life and evolution. Here, we are engineering a yeast strain that can be used to better explore global transcriptional control and its relationship to phenotypic diversity. We are relocating the roughly ~200 DNA binding, RNA polymerase II-associated transcription factors (TFs) in *Saccharomyces cerevisiae* into a single functional module on a neochromosome. The TF neochromosome is being synthesized by assembling ~260 gene-sized pieces of synthetic DNA into a single molecule. We are also constructing TF-free native chromosomes through cloning and reassembly of TF-free natural DNA segments. The outcome of this work will be a cell in which nearly all transcriptional regulation is controlled by a synthetic module on a distinct chromosome. This module will provide a platform for probing minimal sets of TFs required for viability. A minimal TF cohort will then enable explorations of the phenotypic diversity achievable through the reintroduction of accessory TFs.

Conversion of CO_2 Fixated By Cyanobacteria into Carotenoids By *Rhodotorula Mucilaginosa* uanl-001L

Julieta Meléndez, Geovani Rosas-Mejia, Katiushka Arevalo-Niño, Jose Morones-Ramirez

One of the main factors associated with the greenhouse effect intensification is the increase of atmospheric CO_2. Due to human activities, CO_2 emissions have risen up to 68% during the last 21 years. CO_2 is considered a waste from many industrial processes, and it is highly available globally. Cyanobacteria metabolism enables us for the creation of a platform for conversion of CO_2 into added-value carbon-based compounds.

In order to expand the catalog of CO_2 based molecules, in this work we apply a basic principle of synthetic biology, modularization. The first module is CO_2 fixation by the cyanobacteria *Synechocystis* sp. PCC 6803, the second is conversion of the fixated CO_2 into carotenoids by *Rhodotorula*. *Synechocystis* sp. PCC 6803 was cultured in lab-made bioreactors, under strictly autotrophic conditions by supplementing sterilized room air, and using sunlight as light source. Biomass was harvested and implemented as the main component and sole carbon source in the medium for carotenoids production. The yeast *Rhodotorula mucilaginosa* UANL-001L, originally isolated from a local lake in 2016, was selected for the second module due to good performance on growth screening tests. In the proof-of-concept, *R. mucilaginosa* UANL-001L, showed the ability to efficiently metabolize cyanobacterial biomass, achieved similar growth in the tested and in the control medium (YM) and it also showed promising carotenoid production. Thus, carotenoid production was achieved by yeast cultured in media without additional carbon sources other than biomass generated from room CO_2.

Fine Mapping of Functional Genetic Differences between Species By Chromosome Synthesis.

Christopher Ne Ville, Alessandro Luis Venega Coradini, Ian M. Ehrenreich

We recently developed Cloning, Reprogramming, and Assembling Tiled Natural Genomic DNA (CReATiNG), a method for constructing synthetic chromosomes from natural components in yeast. In our initial work, we showed that one application of CReATiNG is to study the genetic basis of trait differences between highly diverged species. However, our initial mapping resolution was coarse, identifying causal loci at the resolution of dozens of genes. Here, we present our efforts to develop an improved version of CReATiNG that can be used to map loci underlying interspecies trait differences to small genomic windows containing only a few genes. To achieve this goal, we are cloning *Saccharomyces cerevisiae* and *S. paradoxus* Chromosome I as 10 evenly sized segments and will assemble all 1,024 possible syntenic combinations of these segments into synthetic chromosomes. Success in our work will produce insights into the genetic basis of trait differences between species and will likely also be beneficial to other applications of CReATiNG.

Exploring Substrate Specificity of Sweets Using a Novel Vacuolar Sugar Transporter Biosensor

Sojeong Gwon

Sugars will eventually be exported transporters (SWEETs) are membrane proteins that mediate cellular sugar uptake and efflux, taking a major part in sugar allocation in plants. Despite extensive research on their structures and molecular functions, it is still unclear how SWEETs recognize their substrates. Using SweetTrac1, a biosensor constructed based on the *Arabidopsis thaliana* SWEET1, we have previously shown the potential of utilizing biosensors and cheminformatics to study the substrate recognition mechanism. Here, we report the new biosensor SweetTrac2, a vacuolar biosensor constructed based on the *Arabidopsis thaliana* SWEET2, to further explore the connection between the protein's substrate specificity. The structural comparison of AtSWEET1 and AtSWEET2 along with each transporter's affinity for different sugars supports the conclusion that substrate recognition by SWEETs is closely connected to non-specific interactions involving key residues in the binding pocket. Furthermore, SweetTrac2 can be an effective tool for monitoring sugar transport at vacuolar membranes that is otherwise challenging to study.

Inducible and Tissue-Specific Synthetic Genetic Circuits for Controlling Plant Development

Bella Archibald, Jennifer A.N. Brophy

Plant synthetic biology holds great promise for combatting the effects of climate change through bioremediation, biomanufacturing, carbon sequestration, and sustainable agriculture, among others. However, the field currently lacks well characterized circuit components necessary for inducible and tissue specific gene expression. My project focuses on creating and comparing inducible circuits for tissue-specific control over lateral root development, which can be modulated to control water use efficiency and to sequester carbon. I am testing different inducible systems and their ability to function in logic gates using transient transformations into *Nicotiana benthamiana*. I am determining the diffusivity of various inducers into *Arabidopsis thaliana* roots to activate tissue-specific gene expression using stable transformations and confocal imaging. In addition, I am performing protoplast assays to identify mutant Auxin Response Factors that can initiate lateral root development independently of native plant mechanisms. Using the results of these experiments, I will build synthetic genetic circuits to activate the tissue-specific process of lateral root development in response to an inducer. The circuit components studied here can be employed to create a variety of additional inducible and tissue-specific plant synthetic circuits.

Engineered Serum Markers for Noninvasive Monitoring of Gene Expression in the Brain

Sangsin Lee, Shirin Nouraein, James J. Kwon, Zhimin Huang, Jerzy Szablowski

Being able to noninvasively monitor gene expression of the brain with a combination of spatial, temporal, cellular, and molecular pathway precision could allow for ongoing gene expression profiling and facilitate understanding of the progression of neurodegenerative disorders[1]. Recently, we developed a paradigm that enables such monitoring of the brain to be performed noninvasively by engineering a class of synthetic biomarkers called **Released Markers of Activity** (RMAs)[2]. In our approach, RMAs are released from the brain parenchyma into the circulatory system by readily crossing the blood-brain barrier (BBB). To achieve this, we leveraged the natural biological process of reverse transcytosis[3], wherein RMAs carry the fragment crystallizable (Fc) protein domain that binds to the neonatal Fc receptor (FcRn) expressed on the BBB endothelial cells and facilitates transcytosis of RMAs from the central nervous system into the bloodstream. By coupling RMAs to specific genes of interest at a local brain site, including subcortical regions, their gene expression levels can be measured and monitored using blood draws. We found that a single intracranial injection of AAV encoding RMA in mice led to over 30,000-fold RMA signal over the background ($p<0.0001$). Additionally, by coupling RMAs to *c-Fos* gene, an immediate early gene (IEG) known to act as a proxy for neuronal activity[4], we observed over a four-fold increase in the RMAs' serum levels after inducing neuronal activation through chemogenetics[5]. Together, RMAs enable longitudinal and noninvasive monitoring of gene expression in the brain with temporal, spatial, cell-type, and molecular specificity, as well as discriminate neuronal activity using blood draws.

References

[1] Farhadi, A., et al. *Nature Materials* **20**, 585-592 (2021). [2] Lee, S., et al. *bioRxiv*, 2022.07.17.500352 (2022). [3] Roopenian, D.C. & Akilesh, S. *Nature reviews immunology* **7**, 715-725 (2007). [4] Sørensen, A.T. et al. *elife* **5**, e13918 (2016). [5] Roth, B.L. *Neuron* **89**, 683-694 (2016).

Engineering Ligand-Specific Biosensors for Neurochemicals and Other Structurally Similar Compounds

Chenggang Xi, Austin Rottinghaus, Matthew Amrofell, Tae Seok Moon

Microbial biosensors have diverse applications in metabolic engineering and medicine (1, 2). Specific and accurate quantification of chemical concentrations allows for adaptive regulation of enzymatic pathways and temporally-precise expression of diagnostic reporters. Although biosensors should differentiate structurally similar ligands with distinct biological functions, such specific sensors are rarely found in nature and challenging to create. Using E. coli Nissle 1917, a generally regarded as safe microbe, we characterized two biosensor systems that promiscuously recognize aromatic amino acids or neurochemicals. To improve the sensors' selectivity and sensitivity, we applied rational protein engineering by identifying and mutagenizing amino acid residues and successfully demonstrated the first ligand-specific biosensors for phenylalanine, tyrosine, phenylethylamine, and tyramine. Additionally, our approach revealed insights into the uncharacterized structure of the FeaR regulator, including critical residues in ligand binding. These results lay the groundwork for developing kinetically-adaptive microbes for diverse applications (3). In this presentation, we will discuss our recent efforts to develop sensors for other structurally similar neurochemicals to enable synthetic neurobiology research and applications.

1. Amrofell, Rottinghaus, and Moon, Curr Opin Biotechnol, 2020. **66**: p. 11-17.
2. Rottinghaus, Amrofell, and Moon, Biotechnology Journal, 2020. **15**: p. 1900319.
3. Rottinghaus, Xi et al. Cell Systems, 2022. **13**: p204-214.e4.

Deciphering the Human Splicing Code with Massively Parallel Reporter Assays

Samantha Koplik, Angela Yu, Madelyn Shelby, Charles Roco, Yue Zhang, Nicholas Bogard, Alex Sabo, Johannes Linder, Georg Seelig

Alternative splicing (AS), a key contributor to proteome diversity, is regulated by a *cis*-regulatory code that is interpreted by *trans*-acting factors with tissue and cell type-specific expression. Mis-regulation of AS has been implicated in a range of diseases. However, a deeper understanding of the sequence determinants of splicing is necessary to engineer such tools to correct aberrant splicing. Massively parallel reporter assays (MPRA) have proven to be a useful tool for simultaneously assaying molecular phenotypes of thousands to millions of sequences in multiple cell types and provide rich datasets for machine learning algorithms. Here, we leverage MPRAs to decipher the sequence and regulatory impacts on AS in high-throughput and build predictive models of AS. We focus on the most prevalent form of AS, exon skipping (ES), and have measured its patterns using over 92,000 minigene splice reporters. The MPRA encompasses short human exons and their flanking introns as well as variants thereof. The MPRA spans 2,159 exons from 1,753 genes and contains 68 sequences from Geuvadis, 6,855 from ExAC, and 84,624 from ClinVar variants. Saturation mutagenesis variations were included in the designed library of splice reporters to provide insight that extends beyond the scale of the human genome, for a more generalized understanding of sequence effects. We have characterized the MPRA library in four human cell lines: HEK293, HepG2, HeLa, and MCF7 and have measured variant effects on ES as well as observed differential splicing patterns between cell lines from the same sequences. These datasets will be used in machine learning models to provide generalizable and predictive models of ES in human cell types. We anticipate that these models can then be used to inform the translational design of programmable and cell type-specific splicing regulatory devices for the treatment of human disease.

A Population-Genomics Approach to Metabolic Engineering of Phenazine Bioproduction in *Pseudomonas Chlororaphis*

Varun Trivedi, Sarah Thorwall, Ian Wheeldon

Metabolic engineering by rational design typically constitutes improvement in biochemical synthesis or other industrially relevant phenotypes by altering one or more biosynthetic pathways in microbial hosts. This approach for selecting genetic targets, although often effective, does not usually target genomic signatures beyond the pathway of interest that could potentially lead to substantial improvements in the phenotype being studied. Recent advances in next-generation sequencing (NGS) technologies have enabled us to sequence whole genomes at low costs. Advancements in bioinformatics have also resulted in increased availability of software tools to analyze NGS reads, among other experimental datasets. These experimental and computational tools could be harnessed to identify novel metabolic engineering targets by conducting experiments and downstream analysis at a pangenome-scale. This work aims to identify non-intuitive genetic targets for improving phenazine biosynthesis in the bacterium *Pseudomonas chlororaphis* by employing a population-genomics approach. We sequenced 34 Pseudomonas isolates using Illumina and Oxford Nanopore technologies and used the NGS reads to assemble whole genomes and construct a pangenome for *P. chlororaphis*. We also characterized three industrially relevant phenotypes – phenazine production, biofilm formation, and growth temperature – for each isolate in two media conditions. Next, we used the whole genome assemblies and the phenazine production data for the 34 isolates to perform a genome-wide association study (GWAS) and identified 330 significant genomic biomarkers for improving bioproduction of various phenazines. Based on a quantitative metric, we elucidated the most influential hits for phenazine production and validated them *in vivo* in the most optimal isolate for phenazine production. Collectively, this work exemplifies the usefulness of adopting a pangenome-wide approach to identify novel metabolic engineering targets for chemical biosynthesis.

Ultra High Throughput Mapping of Gene Circuit Design Spaces in Mammalian Cells

Kshitij Rai, Ronan O'Connell, Trenton Piepergerdes, Bryce Kille, Kristen Curry, Jason Rocks, Todd Treangen, Pankaj Mehta, Caleb Bashor

Despite two decades of gene circuit engineering progress, fine tuning circuit behavior through careful selection of genetic parts remains challenging due to context-specific, often nonlinear interactions with other circuit elements and host machinery. Screening diverse part compositions in high throughput (HT) could dramatically accelerate gene circuit engineering. However, despite advancements in next generation sequencing (NGS) that have enabled HT characterization of libraries of genetic parts, high-depth analysis of libraries of part compositions arranged across multi-kb length scales has not been possible using current NGS platforms. Here, we introduce CLASSIC, a novel genetic screening platform that combines long and short-read NGS platforms to quantitatively assess pooled libraries of DNA constructs of arbitrary length. CLASSIC uses a custom hierarchical cloning scheme to achieve single-pot assembly of constructs linked to short barcodes, and then leverages long-read NGS to construct a circuit-to-barcode lookup table that enables downstream phenotypic assays via short-read NGS. As we demonstrate, CLASSIC achieves complete, balanced assembly of single and multi-gene libraries, enabling phenotypic mapping of genomically integrated single-copy libraries in human cells with remarkable precision and reproducibility. We demonstrate that CLASSIC can scale to massive library sizes by measuring expression profiles of $>10^5$ drug-inducible multi-gene circuit designs in a single experiment; this represents the largest circuit library analyzed to date. Using statistical inference and machine learning approaches, we demonstrate that data obtained with CLASSIC can be used to predictively model the entire circuit design landscape, offering deep insight into the principles underlying circuit design. Our work demonstrates the versatility of CLASSIC for creating "composition-to-function" mappings of synthetic DNA programs, reveal part composability rules and allowing rapid identification of behavior-optimized variants. We foresee that by coupling CLASSIC to ML algorithms, we can dramatically accelerate the pace of genetic circuit engineering, ushering in an era of AI-guided data-driven synthetic biology.

Creating Synthetic Chromosomes from Natural DNA

Alessandro Luis Venega Coradini, Christopher Ne Ville, Zachary Krieger, Joshua Roemer, Cara Hull, Shawn Yang, Daniel Lusk, Ian M. Ehrenreich

De novo chromosome synthesis is costly and time-consuming, limiting its use in research and biotechnology. Building synthetic chromosomes from natural components is an unexplored alternative with many potential applications. In this paper, we report CReATiNG (Cloning, Reprogramming, and Assembling Tiled Natural Genomic DNA), a method for constructing synthetic chromosomes from natural components in yeast. CReATiNG entails cloning segments of natural chromosomes and then programmably assembling them into synthetic chromosomes that can replace the endogenous chromosomes in cells. To illustrate the diverse uses of CReATiNG, we recombined chromosomes between strains and species without mating or meiosis, modified chromosome structure in a programmable manner, and simultaneously deleted 23 non-adjacent genetic elements. CReATiNG enables cost-effective and efficient application of chromosome synthesis to diverse biological problems.

Marginal Specificity in Protein Interactions Constrains Evolution

Dia Ghose, Kaitlyn Przydzial, Emily Mahoney, Amy E. Keating, Michael Laub

The evolution of novel functions in biology relies heavily on gene duplication and divergence, creating large paralogous protein families. Selective pressure to avoid detrimental cross-talk often results in paralogs that exhibit exquisite specificity for their interaction partners. But how robust or sensitive is this specificity to mutation? Here, using deep mutational scanning, we demonstrate that a paralogous family of bacterial signaling proteins exhibits marginal specificity, such that many individual substitutions give rise to substantial cross-talk between normally insulated pathways. Our results indicate that sequence space is locally crowded despite overall sparseness, and we provide evidence that this crowding has constrained the evolution of bacterial signaling proteins. These findings underscore how evolution selects for 'good enough' rather than optimized phenotypes, leading to restrictions on the subsequent evolvability of paralogs.

Protein Engineering Assisted Polyketides Biosynthesis in the Metabolic Network of *Escherichia coli*, Yeast, and Plants

Yu Zhou, Sangdo Yook, Xueyi Xue, Liqing Chen, Yong-Su Jin, Yi Lu

To fulfill the long-term demand for carbon-based products, it is essential to use renewable biomass as feedstocks to generate platform molecules or building blocks for even high value-added products. Triacetic acid lactone (TAL) is a primary example of such platform molecules to be transformed into high-value products such as pharmaceuticals, monomers for polymerization and fuel additives. Several steps of transformation with extra synthetic efforts are needed to convert TAL into such valuable products. To overcome these limitations, we have been working on the direct generation of valuable TAL derivatives and versatile polyketides under mild conditions with controlled selectivity by engineering *Gerbera hybrida* 2-pyrone synthase (2PS), which catalyzes the formation of TAL naturally, and reprogramming 2PS's biocatalytic pathway into the metabolic network to produce high-value products directly in living cells. Toward this goal, we have generated one 2PS mutant that displayed 12-fold higher yield, 310-fold larger k_{cat} and 70-fold higher catalytic efficiency of triketide styrylpyrone production than wild type 2PS. The chemoselectivity can be switched by protein engineering to generate diketide product or tetraketide product which wild-type 2PS fails to produce. Through metabolic engineering, different 2-pyrone products can be directly generated by *E. coli* and *Yarrowia lipolytica* by feeding easily available carboxylic acids with yields up to 17 mg/L culture. An *in vivo* biocatalytic cascade has been set up to apply directed evolution to enhance the reactivity of mutated 2PS, which improved the titer to over 200 mg/L in culture tubes. Such engineered 2PS-catalyzed novel reactivity has been further implemented in the metabolic network of *Nicotiana benthamiana* to directly produce various psychoactive kavalactones. Empowered by protein engineering and synthetic biology, one enzyme scaffold is designed to synthesize versatile polyketide products in different host systems, which not only provides direct TAL functionalization methods but also offers insights into polyketide synthase mechanistic study.

Identification of Selective Glycosyltransferase Inhibitor Molecules Targeting Cell-Surface Fucosylation By High Throughput Methods

Dulce María Valdez López, Pierre-André Gilormini, David J. Vocadlo, David H. Kwan

In cancers, increased fucosylation of cell surface glycans is a trademark of malignant cell transformation and contributes to many aberrant cellular events during cancer development. Alterations in the expression of fucosylated oligosaccharides is mainly attributed to the upregulation of specific fucosyltransferase enzymes (FUTs), which construct fucosylated glycans using GDP-fucose as a substrate. FUT5 is an a1-3-fucosyltransferase that acts to form sialyl Lewis^X, a member of the Lewis antigen family that is abnormally abundant in numerous cancers and is associated with advanced tumor grade poor prognosis due to its capacity to interact with E- and P- selectin molecules expressed by the endothelium, thus aiding in metastasis by direct binding of cancer cells.

To identify inhibitors of the FUT5 enzyme as potential cancer therapeutics, we have developed a coupled high-throughput assay that makes use of 3'sialyl-N-acetyllactosamine-resorufin (3'SLN-Res), a fluorogenically labeled sialyl Lewis^X precursor. When non-fucosylated, this synthetic substrate is recognized and hydrolyzed by specific glycoside hydrolase enzymes to release fluorescent resorufin, yet if the probe is fucosylated by FUT5 prior to treatment with glycosidases, these cannot act on the substrate and no fluorescent signal is produced. This assay is suitable to measure activity and inhibition of FUT5 by drug-like small molecule compounds because blocking fucosylation allows glycosidase-catalyzed hydrolysis of the labeled oligosaccharide to produce a fluorescent signal. Employing this assay, we screened against a targeted library of ~1000 drug-like small molecule compounds in a high-throughput setting and identified several hits, demonstrating that our approach can be used to find potent FUT inhibitors involved in the formation of sialyl Lewis^X.

Genoscaler™: A Next-Generation High Throughput Genome Engineering Platform for Rapid Product Development and Commercialization

Richard Fox

New genome engineering approaches for the efficient development of sustainable, cost-effective biomanufacturing solutions are sorely needed. To address this acute demand, Inscripta has developed GenoScaler™, a proprietary, next-generation, first of its kind, rapid microbial genome engineering platform. GenoScaler™ is a scalable, cost-effective, high-throughput CRISPR-based genome engineering technology stack that can rapidly design, build, and test 10,000+ precision edits across *E. coli* and *S. cerevisiae* genomes in both single and multiplexed (combinatorial) formats. We demonstrate massively parallel, highly efficient editing with a variety of edit types that have been challenging or impossible to make at scale using traditional CRISPR methods. We have applied GenoScaler™ to rapidly develop manifold improved strains across a variety of applications for endogenous & heterologous small molecule and protein production. Along with integrated, state-of-the-art methods for high-throughput phenotyping, smart automation, scale-up, informatics, and machine learning, the platform is broadly applicable and can accelerate the process of synthetic biology product development by orders of magnitude via unprecedented gains in speed, efficiency, and performance compared to traditional methods for strain engineering. The platform allows Inscripta and its partners wide access to the bioeconomy, from concept through to scale-up and commercial production and ingredient supply using the newly developed principles of Lean Bioengineering™.

Conditional Transgene Expression Triggered By Specific mRNA Sequence

Lara Selles, Mayumi Suzuki

Regulation of transgene expression in response to specific mRNAs enables context-dependent gene expression that will benefit gene therapy and metabolic engineering. Group I intron ribozymes were the first discovered catalytic RNA molecules. They can self-splice without protein factors, only requiring the presence of GTP and Mg^{2+}. Here, we present a toolkit of several synthetic designs that enable regulation of gene expression at the mRNA level in response to the presence or absence of specific nucleic acid sequences. The designs are based on split *Tetrahymena thermophila* group I intron fragments fused to mRNA precursors. The non-functional ribozyme fragments interact and reconstitute a fully functional group I intron in the presence of a specific trigger oligonucleotide or mRNA. Subsequent self-splicing results in the ligation of the two fragments of the split transgene, delivering a functional mRNA ready to be expressed. Several split sites appropriate for the design of such bimolecular constructs have been identified, which also allow for modifications in the group I intron sequence to make it more compact or to make it responsive to different trigger sequences. Therefore, the modularity of the designs facilitates adaptations to different transgenes and trigger sequences. Such properties make this strategy suitable for conditional expression of therapeutic genes triggered by an endogenous mRNA characteristic of abnormal or pathological cells.

Computational Design and Discovery of Multi-Functional Hydrolase for Detoxifying Cocaine and Its Toxic Metabolites in the Concurrent Use of Cocaine and Alcohol

Fang Zheng, Chang-Guo Zhan

Cocaine abuse is a major medical and health problem. There is no FDA-approved medication for the treatment of cocaine abuse. Statistical data show that 92% of cocaine users also consume alcohol. The risk of immediate death is 18 - 25 times greater for cocaine co-ingested with alcohol than for cocaine alone. Alcohol can react with cocaine to result in a series of toxic compounds in body including cocaine, cocaethylene, norcocaine, and norcocaethylene. Notably, metabolites cocaethylene, norcocaine, and norcocaethylene are all more toxic than cocaine itself. Through a combined use of our "virtual screening of transition states" computational protocol and artificial intelligence (AI) methods, a novel approach has been used to design butyrylcholinesterase (BChE) mutants as multiple-functional cocaine hydrolases (mfCocHs) for treatment of the toxicity caused by concurrent use of cocaine and alcohol. In comparison with the kinetic parameters of *native human BChE*, the most effective mfCocH has at least ~1000-fold improved catalytic efficiency against three of the toxic substrates (cocaine, norcocaine, and cocaethylene) and ~68-fold improved catalytic efficiency against norcocaethylene. *In vivo* studies in rats revealed that the mfCocH can effectively hydrolyze cocaine and all the three toxic metabolites, thus quickly detoxifying cocaine and its three toxic metabolites in plasma. This was demonstrated by our rescue experiment in rats co-administrated 1 g/kg alcohol (IP) and a lethal dose (180 mg/kg, IP, LD_{100}) of cocaine.

Scalable, Continuous Directed Evolution of Antibody-Based GPCR Ligands

Adam Sanford, Arjun Ravikumar, Ahmad S. Khalil

G protein-coupled receptors (GPCRs) are one of the most common signaling structures in the human body and control numerous physiological processes including vision, olfaction, neurotransmission, nutrient metabolism, and body homeostasis. Accordingly, 1/3 of all FDA approved drugs target GPCRs; yet, more than 75% of all non-olfactory GPCRs, including many well-validated drug targets, remain undrugged. This drug discovery gap is largely due to three reasons: off-target drug action at GPCRs closely related to the target, on-target effects at undesired tissues, and poor drug pharmacokinetics. Antibody-based GPCR therapeutics have been long sought to overcome these challenges with their exquisite specificity and robust pharmacokinetics, but our current antibody discovery methodologies struggle to uncover the rare antibody variants capable of precisely modulating GPCRs to elicit desired downstream signals. Here we propose leveraging directed evolution to discover functional antibodies for difficult GPCR targets entirely within engineered cells, thereby avoiding the onerous receptor purification and structure determination or animal immunization that is typically required for GPCR antibody drug discovery. To achieve this, we functionally reconstituted several human GPCRs into *Saccharomyces cerevisiae* and created genetic circuits that couple receptor activation to expression of growth-selectable markers. We then co-express a library of antibodies that are tethered proximal to the GPCR, where they can activate receptor signaling, allowing us to select for library members that modulate the GPCR simply by growing the cells in selective media. By combining continuous directed evolution techniques with tunable selection stringencies, we can mature initial library hits for potency and receptor selectivity. This platform provides a novel, flexible approach for discovering candidate therapeutic biologics against difficult-to-drug yet important GPCR targets.

High Complexity, One Pot Golden Gate Assembly Systems for the Rapid Engineering of Small Genomes

Oluwapelumi Adeyemi, Katharina Bilotti, Barbra D. Ferrell, Katalin Foldes, Sarah Keep, Rachel A. Keown, Shawn W. Polson, Vladimir Potapov, John Pryor, Andrew Sikkema, Kasra Tabatabaei, Erica Bickerton, K. Eric Wommack, Gregory J. S. Lohman

We have previously demonstrated using Data-Optimized Assembly Design (DAD) for the selection of fusion sites permits high complexity golden gate assembly of dozens of parts to be assembled in a single reaction. We have applied these principles to develop systems for assembling T7-like bacteriophage genomes and genomic cDNA for Avian Infectious Bronchitis Virus (IBV). These systems permit assembly of the full targets in a single GGA step from 12-36 parts. Parts are rationally designed to isolate many ORFs and produce fragments that are, with few exceptions, readily propagated and manipulated in traditional E. coli plasmid systems. The design thus supports easy viral mutagenesis via classic molecular biology methods as well as whole gene swaps and produce final assembly products that are compatible with classic methods for rebooting the viruses. The methods highlighted here are applicable to the design of flexible and rapid assembly strategies for other DNA targets of less than ~50kb, potentially accelerating the study of viral biology, engineering of functional phages, and development of attenuated strains for vaccine development.

Usurpation: Utilizing a Novel Suppression System to Introduce Non Canonical Parts into Translation

Martin Edelmann

Translation is a central process in bacterial life and beyond. In all bacteria the key players of this process are highly conserved in their structure. Here we created a novel genetic system, which metabolically couples the growth of *E. coli* with the *in vivo* activity of foreign non-canonical components in translation. Due to the extreme sensitivity and the high safety of the system, it is ideal to be applied in long term evolution experiments. In those we are forcing *E. coli* to evolve to accept the foreign components in the framework of the cellular translation apparatus. This enables us to answer questions such as: Is it possible for bacterial translation to look differently then we see it in nature or what does need to change in a cell to enable the utilization of a novel kind of molecule in translation.

In a first application for the system, we focused on the tRNA. All tRNAs in *E. coli* have the typical cloverleaf secondary structure, which leads to a L-shaped 3D fold. Using the novel system, we successfully evolved *E. coli* to utilize a different fold in translation – tRNAs that completely lack the D-arm. To our surprise the complete translation machinery accepted a D-arm-less tRNA fold *ab initio* and the only adaptation required was in the (t)RNA degradation machinery. We confirmed our results genetically and *in vitro* by utilizing the minimized tRNA in a reconstituted *in vitro* translation system. This is the first reported synthetic tRNA with a non-canonical fold that was introduced into the translation process of a living organism. Our work suggests that life allows for more flexibility in translation as we currently observe it in nature.

Machine-Learning Guided Design of Cell-Specific Adeno-Associated Viral Vectors (AAVs) in Human Central Nervous System (CNS)

Danqing Zhu, David Brookes, Akosua Busia, Jennifer Listgarten, David Schaffer

The CNS comprises a multitude of cell types with diverse functionality and specialization. Dysregulation of neuronal or glial populations has been implicated in multiple disorders. AAVs hold tremendous promise as gene delivery vectors to treat such conditions given their reasonable starting efficiency and safety profile. However, such strategies remain challenging due to difficulties in efficient and targeted delivery to specific cell populations. Directed evolution can achieve improvements in AAV performance without pre-existing knowledge of virus-cell interactions. Here we apply directed evolution technologies guided by machine-learning (ML) design for developing efficient AAV vectors that target specific CNS cell populations in human brain tissues.

To optimize our library design, we first developed a ML-based framework to design an AAV5-capsid library. Using the trained predictive model, we designed individual libraries that lie on an optimal curve that trades off predicted enrichment with maximal sequence diversity. We then synthesized two of the libraries and showed that our ML-library achieves 5-fold higher packaging fitness than the state-of-the-art library, with approximately 10-fold more successful variants after primary brain infection. We transfected our synthetic ML-designed library into packaging cell lines and administered to primary brain slices. Functional selective pressure was imposed by harvesting glial- and neuronal-only cells after infection. Using PCR-based recovery of variants with deep sequencing, we identified variants that successfully transduce human glial or neuronal cells specifically.

In summary, our approach offers a roadmap for designing an AAV library that broadly improves packaging functionality while retaining high diversity. This approach can be harnessed to optimize libraries for downstream selection for properties of therapeutic relevance, including production, infectivity, specificity, and others. We also anticipate that such ML-guided design of AAV vectors will have broad utility for the development and selection of novel variants targeting different cells and tissues for applications in the near future.

Appraise: Fast, Accurate Ranking of Engineered Proteins By Receptor Binding Propensity Using Structure Modeling

Xiaozhe Ding, Xinhong Chen, Erin E. Sullivan, Timothy F. Shay, Viviana Gradinaru

Deep learning-based methods for protein structure prediction have achieved unprecedented accuracy. However, the power of these tools to guide the engineering of protein-based therapeutics remains limited due to a gap between the ability to predict the structures of candidate proteins and the ability to assess which of those proteins are most likely to bind to a target receptor. Here we bridge this gap by introducing Automated Pairwise Peptide-Receptor AnalysIs for Screening Engineered proteins (APPRAISE), a method for predicting the receptor binding propensity of engineered proteins. After generating models of engineered proteins competing for binding to a target using an established structure-prediction tool such as AlphaFold-Multimer or ESMFold, APPRAISE performs a rapid (under 1 CPU second per model) scoring analysis that takes into account biophysical and geometrical constraints. As a proof-of-concept, we demonstrate that APPRAISE can accurately classify receptor-dependent vs. receptor-independent engineered adeno-associated viral vectors, as well as diverse classes of engineered proteins such as miniproteins targeting the SARS-CoV-2 spike protein, nanobodies targeting a G-protein-coupled receptor, and peptides that specifically bind to transferrin receptor and PD-L1. APPRAISE is made accessible through a web-based notebook interface using Google Colaboratory (https://tiny.cc/APPRAISE). With its high accuracy, interpretability, and generalizability, APPRAISE has the potential to expand the utility of current structural prediction methods and accelerate protein engineering for biomedical applications.

Picoshells: A Tool for Screening Millions of Protein Variants in Application-Relevant Environments

Mark van Zee, Rajesh Ghosh, Timothy Vernon, Issei Yamaguchi, Cayden Williamson, Diego Salgues, William Vo, Takuya Terai, Robert E. Campbell, Dino Di Carlo

Current protein screening technologies represent a major bottleneck in our ability to evolve and engineer new proteins because of their low throughput and/or inability to place proteins into their native structure or solutions relevant for their end application. To overcome these limitations, we have developed PicoShell technology, a novel protein screening tool that allows scientists to screen millions of variants, easily swap out the soluble environment the proteins are exposed to, and place proteins in solutions that are useful for their end application (e.g. serum, buffer, etc). PicoShells are spherical microparticles with a hollow center and an outer porous shell made of polyethylene glycol that can be fabricated at a rate of 2000 particles per second. Cells are placed into the hollow center where they can either secrete or be lysed to release the desired proteins. The Picoshells pore size can be engineered to be anywhere between 15 and 250 kDa such that proteins and plasmids of interest retained within the hollow center. PicoShells can also be engineered to contain capture motifs that can bind to released proteins. Due to the free diffusion of small molecules across the shell, the interior solution where the proteins are located can be changed by centrifuging down the PicoShells and exchanging the solution the particles are suspended in. This feature enables assays to select proteins that respond to changes in analyte concentrations from other proteins that are less- or non-responsive and place proteins in any desired solution. PicoShells can be coupled with standard fluorescence-activated cell sorting (FACS), which enables us to screen millions of protein variants with throughputs of >50 variants per second. We expect that the PicoShell technology will dramatically accelerate the process of protein engineering and increase the range of selection criteria that can be used to guide protein directed evolution efforts.

Characterizing the Functional Repertoire of Protein Dimerization Networks

Jacob Parres-Gold, Matthew Levine, Benjamin Emert, Andrew Stuart, Michael Elowitz

Cells use circuits of interacting proteins to interpret and respond to signals from other cells, the environment, and their own cell state. Many natural biochemical circuits employ families of protein variants that interact with one another in a many-to-many fashion; for example, the bZIP, bHLH, and nuclear receptor transcription factor families comprise dozens of protein monomers that can dimerize in many possible combinations before binding to DNA. While previous work has demonstrated that the responses of such dimerization networks are complex, little is known about the overarching range of computations that dimerization networks can perform – their "expressivity" – and what advantages those computations offer for cellular information processing. In this work, we use both random parameter screens and optimization trials to characterize the expressivity of combinatorial dimerization networks, treating network proteins (monomers) as inputs and individual dimers are outputs. We find that networks of just three or four monomers can compute complex, non-monotonic functions on multiple inputs and that increasing network size allows new classes of functions to be performed. Further, we demonstrate that the same network of proteins can perform different functions if expressed at different abundances, offering a potential mechanism for cell type-specific responses. Thus, dimerization networks can perform complex and cell type-specific computations using small sets of simple components that can be readily generated over evolution through gene duplication and divergence.

Engineering Orthogonal Phase-Separated Compartments Using Modular RNA Motifs

Anli Tang, Shiyi Li, Jaimie Marie Stewart, Melissa Klocke, Paul W. K. Rothemund, Elisa Franco

The process of liquid-liquid phase separation of biomolecules such as proteins and nucleic acids drives the formation of a variety of cellular organelles that carry out important physiological functions. By harnessing phase separation through the use of engineered molecules it is becoming possible to build artificial organelles that can spatially organize cells and introduce new functionalities. Here we report the design and synthesis of orthogonal RNA-only condensates from folded single-stranded RNA monomers that assemble based on HIV-type kissing loop (KL) interactions. These RNA monomers, referred to as RNA nanostars, consist of multiple stem-loop structures with KL sequences located in each distal loop. We investigate the effects of the intrinsic factors, such as stem sequence, stem length and KL sequence, as well as the extrinsic factors, like temperature and ionic conditions. We demonstrate that the RNA nanostars can form distinct and immiscible compartments using multicolor fluorogenic RNA aptamers. Furthermore, we validate that these RNA nanostars can cotranscriptionally form condensates at 37C *in vitro*. Our work establishes strategies for developing a library of programmable and orthogonal RNA condensates, with potential applications in engineering modular condensates that can control cellular processes within living cells.

Sugar, Spice, and Everything Nice: Directed Evolution of Glucose-Dependent Enzymes Via Growth Complementation of Glucose Non-Utilizing *E. coli*

Chiagoziem Ngwadom, Caroline M. Ajo-Franklin

Why should we as do work bacteria are more than happy to do for us? How can I "search" a question and encourage microorganisms to give me the answers they have available? Soluble pyrroloquinoline quinone (PQQ)-glucose dehydrogenase (PQQ-GDH) is an exceptionally active sugar-converting biocatalyst with catalytic turnover independent of oxygen and in broad pH and temperature ranges. PQQ-GDH native sequence, structure, and function have been elucidated in recent years. This has allowed PQQ-GDH to be modified for biosensing and bioenergetics, particularly as switches that can sense and respond to molecular inputs. Recently, we have explored the structure-function-sequence of this enzyme by its ability to tolerate octapeptide insertions. By uncovering the regions favorable to domain insertion, we have also engineered PQQ-GDH to be a simple, two-component switch capable of changing the current signal in the presence of cancer biomarkers such as estrogen and its analogs. Still, the screening technique currently used is elementary, and the utility of trying to enable functionalities via semi-rational or rational design techniques is limited. *Escherichia coli* (*E. coli*) can metabolize many carbohydrate sources, though it prefers D-glucose as its primary carbon source. A series of mutations to inhibit glucose transport and utilization has allowed us to use glucose-negative *E. coli* strains for the functional complementation of PQQ-GDH, thus using directed evolution for the selection of favorable PQQ-GDH mutants. The impact of this work has allowed for fast and dynamic selection of glucose-dependent oxidoreductases, thus advancing the engineering of enzymes for bioelectrocatalysis, health diagnostics, and genetic circuits.

Integrating New Diversity Technologies into the Development Platform to Produce High-Performing Industrial Enzymes and Biomolecules

Michael Liszka

For over 35 years, BASF has produced biobased products for food, health, pharmaceutical, crop protection and chemical industries by fermentation and biocatalysis. BASF possesses end to end core technologies for bioproduct development spanning the discovery of novel genes, gene product engineering and high-throughput screening, and strain and process development. Industrial enzymes, in particular, have the potential to drive sustainable solutions by increasing efficiency across food and traditional industries and reducing carbon dioxide emissions. We continue to build and maintain intimate application knowledge in our strategic markets to deliver relevant products to our customers. Our presentation highlights several case studies where BASF colleagues have applied these techniques to optimize enzyme properties including thermal tolerance, substrate specificity, and expression. We'll demonstrate the discovery and development of new industrial enzymes and other bioproducts using iterative and computer-aided design techniques, including how data driven design cycles can improve an enzyme's biophysical properties and performance. In addition, we'll illustrate how machine learning is starting to be applied towards developing new amino acid sequences to drive enzyme production. Finally we will discuss how BASF is pivoting to also help accelerate and mature technology of collaborators and customers in order to support a growing Bioeconomy and develop sustainable solutions for multiple industries.

Advances in Biomolecular Systems Engineering

Raja Selvakumar

Biomolecular systems engineering represents the confluence of biological, chemical, and electrical engineering with the explicit intent of engineering intelligent biological systems. This emerging field is distinguished from synthetic biology given the additional emphasis on protein engineering. Fundamentally redesigning protein-DNA interactions unlocks higher performance of intelligent chassis. The bespoke systems will be capable of advanced decision-making, programmable memory, and targeted cell to cell communication. These three foundational tenets enable the development of chassis-agnostic living therapeutics and biomanufacturing applications.

This work broadly applies to Cell Manufacturing and Drug Delivery. For Cell Manufacturing, a high-performance industry relevant microbe with programmable function is desired. Advanced decision-making and programmable memory have been demonstrated in E. coli with ongoing research for the same in gram-positive B. subtilis. Future work will redefine the biotechnology paradigm by enabling targeted cell-cell communication with high precision population control. For Drug Delivery, a non-pathogenic human-compatible microbe with programmable function is desired. Advanced decision-making has been demonstrated in B. supp with ongoing research for memory and added functionality in E. coli Nissle 1917.

Dynamic Division of Labor Facilitated By Horizontal Gene Transfer

Grayson Hamrick, Hye-In Son, Lingchong You

Engineered microbes have shown great potential for applications such as bioprocessing, environmental remediation, waste management, and biosensing. However, the complexity of the functions performed by these microbes is fundamentally limited by the metabolic burden imposed on them by their heterologous gene circuits. To overcome this limitation, synthetic biologists have explored division of labor (DOL) strategies designed to distribute the burden of a taxing process across a consortium of microbes. In DOL, tasks are distributed between several microbial strains, as opposed to burdening a single microbe with all parts of the desired process. Simple DOL is subject to a variety of design difficulties–such as population ratio instability due to competition–and is therefore difficult to implement. We propose a method of DOL mediated by horizontal gene transfer (HGT) to improve both the stability of the functional microbial population and the throughput of burdensome metabolic pathways. Through mathematical modeling and numerical simulations, we show that HGT has the potential to buffer a productive bacterial consortium from the destructive effects of high burden by maintaining functional population ratio and productive stability.

Engineering Antiviral Immune-like Systems for Autonomous Virus Detection and Inhibition in Mice

Yidan Wang

The COVID-19 pandemic has demonstrated that viral diseases represent an enormous public health and economic threat to mankind and that individuals with compromised immune systems are at greater risk of complications and death from viral diseases. The development of broad-spectrum antivirals is an important part of pandemic preparedness. Here, we have engineered a series of designer cells which we term autonomous, intelligent, virus-inducible immune-like (ALICE) cells as sense-and-destroy antiviral system. After developing a destabilised STING-based sensor to detect viruses from seven different genera, we have used a synthetic signal transduction system to link viral detection to the expression of multiple antiviral effector molecules, including antiviral cytokines, a CRISPR-Cas9 module for viral degradation and the secretion of a neutralising antibody. We have performed a proof-of-concept study using multiple iterations of our ALICE system *in vitro*, followed by *in vivo* functionality testing in mice. We show that dual output $ALICE_{SaCas9+Ab}$ system delivered by an AAV-vector inhibited viral infection in herpetic simplex keratitis (HSK) mouse model. Our work demonstrates that viral detection and antiviral countermeasures can be paired for intelligent sense-and-destroy applications as a flexible and innovative method against virus infection.

Engineering Artificial Phosphorylation Networks in Mammalian Cells

Xiaoyu Yang, Jason Rocks, Kaiyi Jiang, Kshitij Rai, AJ Walters, Pankaj Mehta, James J. Collins, Nichole Daringer, Caleb Bashor

Cells universally use signaling networks to perceive and compute changes to their external environment. While phosphorylation is the dominant molecular currency for transmitting information within these networks, engineering synthetic phosphorylation-based signaling circuitry in mammalian cells has been challenging due to a lack of robust parts and design schema. Here we identify the enzymatic push-pull motif—in which two opposing catalytic activities control the covalent modification of a target protein—as the fundamental unit for constructing synthetic phosphorylation networks. In this work, we develop a parts toolkit for engineering with kinase/phosphatase push-pull devices and show that the push-pulls can be used to construct synthetic phosphorylation-based signaling networks in HEK293T cells. By interlinking and tuning push-pulls, we can create diverse circuit topologies: when connected upstream to diverse synthetic receptors classes, inducible push-pulls can sense and respond on a fast timescale to changes in extracellular ligand concentration, while coupling push-pulls to downstream outputs can enable regulation of diverse cellular behaviors, including reversible formation of liquid phase condensates, activation of gene expression, and rapid induction of secretion. To enable predictive engineering of phospho-signaling circuits, we developed a novel thermo-inspired modeling framework capable of describing non-equilibrium network behavior. We show this framework can be used to identify circuit designs that demonstrate specific target behaviors from amongst a large design space. Finally, we showcase the translational utility of our system by implementing a TNF-a sensing and IL-10 secreting network in engineered cells for alleviating inflammation, demonstrating the potential for our synthetic phosphorylation-based pathways to be used in cell therapy applications. Our work introduces a broadly applicable synthetic biology framework for constructing post-translational regulatory networks in mammalian cells that should facilitate the development of therapeutic applications by enabling model-driven, user-defined programming of a cell's ability to sense and respond to molecular cues in its external environment.

Constructing Synthetic Gene Circuit Using Ultra-High Affinity Protein Degradation Tag in Escherichia coli

Sudeshna Roy, Prajakta Jadhav, Nicholas Butzin

The ClpXP protease complex is responsible for the degradation of many intracellular proteins but is maintained at low levels in *E. coli* to avoid non-specific degradation. The protease can be a rate-limiting step in the process of protein degradation leading to proteolytic queueing, proteins wait in line to be degraded. Synthetic biologists have leveraged proteolytic queueing to design robust synthetic circuits by tagging proteins with SsrA tags, an 11 amino acid sequence recognized by the ClpXP protease. However, the component of ClpXP complex responsible for queueing was unknown. Previous work had demonstrated the binding site of each complex to the SsrA tag. We created a library of SsrA tag variants and overexpressed each protein in the complex, *in vivo*. Based on the degradation pattern of the proteins with SsrA variants and overexpression of each individual component of the protease complex, we determined that the ClpP (protease) and SspB (adapter ATP-independent chaperone) are not responsible for queueing, but provide evidence that the queue is at ClpX (ATP-dependent chaperone). We, further, measured the binding affinity of the SsrA variant tags and observed that LAALAA had a higher binding affinity to ClpXP than the original SsrA tag, AANDENYALAA. To understand the functionality of the ultra-high affinity tag in a dynamic-synthetic circuit, we modified the dual feedback oscillator by replacing the original SsrA to the SsrA variant tag, LAALAA. Analysis of the frequency of oscillations in the modified dual feedback oscillator using microfluidics exhibited altered oscillations. Like the original dual feedback oscillator, thousands of cells with the modified dual feedback oscillator could be synchronized by entrainment using an external signal in a microfluidics device. These results demonstrate the power of deconstructing the mechanism of a natural system (basic research) and how its knowledge can be used to develop new, complex synthetic systems in bacteria.

Leveraging the Carbon Storage Regulatory Network to engineer Post-Transcriptional Gene Circuits

Trevor Simmons, Gina Partipilo, Anna Stankes, Darian Chiu, Benjamin Keitz, Lydia Contreras

Historically, bacterial gene circuits have relied upon transcriptional regulatory elements; including synthetic promoter libraries, engineered transcription factors, and CRISPR-based systems. However, there is now a push to develop similar genetic circuits using post-transcriptional regulatory systems, as they maximize system response time and minimize cellular resource competition. Yet, there are few post-transcriptional systems that offer dynamic control over multiple heterologous targets, and bi-directional regulatory outcomes. With this in mind, the Carbon Storage Regulatory (Csr) Network of *Escherichia coli* serves as an ideal system for engineering to address current limitations. The Csr Network natively activates and represses hundreds of mRNA targets, and rapidly responds to environmental signals. Herein, we leverage the native Csr Network of *Escherichia coli* to create semi-synthetic Buffer gate, and NOT gates for heterologous target regulation. We expand upon our initial designs to create a genetic toolbox for precise titration of gene expression over a 10-fold range. As a downstream proof-of-concept, we apply this toolbox to a synthetic mevalonate operon. Using these semi-synthetic circuits, we demonstrate a five-fold increase in mevalonate production relative to a transcriptionally controlled operon. Lastly, as the Csr Network is conserved throughout the *Gammaproteobacteria* class, we demonstrate these genetic circuits can be transferred across multiple organisms by leveraging the native components of the Csr Network within each organism.

Developing an Embedded Plate-Reader Device for Automated Optogenetic Dynamical Control

Krishna Pochana, Fatemeh Aghlmand, Richard Murray, Azita Emami, Chelsea Hu

Optogenetics is a powerful tool that allows gene expression to be controlled by light, holding immense potential in enabling feedback control between engineered biological systems and external controllers in synthetic biology. However, testing such biological systems during the engineering phase can be costly, either in terms of developing machinery to automatically test cultures or the labor needed to present signal patterns and manually obtain response measurements. To address this issue, we present an integrated electronic device for optogenetic dynamics testing with up to 16 replicates that fit within the footprint of a standard 96-well plate. The device contains a microcontroller, battery, and 16 control sites, with each site being equal in dimension to a 96-well-plate well and featuring three controllable LEDs emitting in the 620-625nm, 522-525nm, and 465-467nm ranges. Optically transparent and replaceable inserts containing cultures are placed in each control site, allowing the device to be reused by replacing the inserts. A Bluetooth link from the internal microcontroller to an external computer allows LED wavelength and intensity to be controlled based on plate reader data for each control site. In this work, we detail the design of the device and inserts using two materials: PDMS, a standard material, and 3D printable resin, a novel, and cost-effective option. Finally, we demonstrate the device's efficacy using E. Coli cultures engineered for optically induced sfGFP production and show controlled sfGFP production through fixed-timing control.

Engineering Intelligent *Bacillus Subtilis* Chassis Cells

Sumin Lee

Bacillus Subtilis (B.subtilis) has been a popular workhorse in biomanufacturing and more recently in synthetic biology. One of the key features of *B.subtilis* is the chassis cell's efficient protein secretion system, enabling the synthesis and secretion of valuable products – *e.g.*, medicinal proteins and industrial enzymes. This prevalence of *B.subtilis* makes it an ideal candidate to engineer as a programmable chassis cell for biomanufacturing. Here we demonstrate a design, build, and test workflow for conferring decision-making (Transcriptional Programming) in *B.subtilis* through the pairing of synthetic transcription factors with cognate inducible synthetic promoters. This collection of synthetic transcription factors and cognate synthetic promoters can be paired to form objective Boolean logical operations for gene regulation. The work presented here shows the performance of the fundamental single-input single-output logic gates responsive to a pool of chemical signals in the *B.subtilis* chassis cell. This set of basic logical operations represent the building blocks for subsequent combinatorial programs that can be expanded to multiple-input single-output systems. This work establishes the fundamental toolkit for programming gene expression in *Bacillus Subtilis* for use in intelligent biomanufacturing.

Unleashing the Power of Light: Regulation of Gene Expression in a Nanobiosystem Featuring Metallic Nanoparticles and Genetically Modified *E. coli*

Jose Morones-Ramirez

In recent years, genetic engineering and synthetic biology have led to a significant breakthrough in our ability to produce complex biological products on a large scale, utilizing genetically-modified microorganisms equipped with finely-tuned genetic circuitry. Despite numerous successes, there are still a number of challenges associated with developing bioproduction processes involving microorganisms with multiple or complex gene circuits. In particular, external gene expression regulation methods that provide spatial-temporal control and scalability, while inducing minimal permanent or irreversible system-wide effects, are difficult to develop. Stimuli such as pH and chemical inducers can cause permanent, irreversible changes, whereas temperature, ultrasound, magnetic fields, and light can cause partially or fully reversible changes in the culture media conditions.

Opto-control of gene expression is a non-invasive external trigger that satisfies most of the desired characteristics of an external control system. However, its disadvantage lies in the need to design biological photoreceptors that respond to a different wavelength for each bioprocess that needs to be controlled or regulated in the microorganism. To overcome this challenge, we propose the use of biocompatible metallic nanoparticles as external controllers of gene expression, owing to their ability to convert light into heat and the versatility of nanotechnology to create a wide range of nanostructures that can absorb light at different wavelengths and induce plasmonic photothermal heating.

Our research has led to the development of a nanobiosystem that can be opto-thermally triggered using LED light, consisting of biocompatible gold nanoparticles and a genetically modified *E. coli* carrying a plasmid that allows mCherry fluorescent protein production at 37°C in response to an RNA thermometer.

In conclusion, our LED-triggered photothermal protein production system offers a novel, cost-effective, scalable, and non-destructive method that can contribute to the advancement of bioprocess production systems.

A Platform for Distributed Production of Synthetic Nitrated Proteins in Live Bacteria

Neil Butler, Aditya Kunjapur

The incorporation of the non-standard amino acid (nsAA) *para*-nitro-L-phenylalanine (pN-Phe) within proteins introduces a chemical functionality that exhibits distinct electrostatic properties from the standard twenty amino acids. Proteins that contain pN-Phe have been used for diverse applications including the termination of immune self-tolerance. However, the promise of nitrated proteins for roles such as increased protein immunogenicity is limited by the requirement for provision of chemically synthesized pN-Phe. Here, we report the construction of a live bacterial producer of synthetic nitrated proteins by coupling metabolic engineering and genetic code expansion. We designed a metabolic pathway to achieve first-time biosynthesis of pN-Phe in *Escherichia coli*, using a diiron monooxygenase-type *N*-oxygenase to biosynthesize the nitroaryl chemical functional group from an aryl amine precursor. To improve the relevance of pN-Phe biosynthesis for eventual incorporation within protein sequences, we pursued several strategies to improve titer, rate, and selectivity. By expressing a previously uncharacterized *N*-oxygenase in an *E. coli* strain that was engineered for greater shikimate pathway flux, we achieved pN-Phe titers of 820 ± 130 μM using MOPS EZ Rich media. After we identified a variant of the *Methanocaldococcus jannaschii* tyrosyl-tRNA synthetase and tRNA pair that was selective for pN-Phe rather than the *para*-amino-L-phenylalanine precursor metabolite, we coupled pN-Phe biosynthesis and incorporation in a single strain. The resulting strain generated reporter protein titers of 0.53 mg/L with pN-Phe containing protein as the primary product. Overall, our study has created a foundational technology platform towards distributed and autonomous production of nitrated proteins.

Engineered *Bacillus Subtilis* Capable of Epithelial Cell Invasion

Emily Greeson, Cody Madsen, Evran E Ural, Ashley Makela, Victoria A Toomajian, Christopher H. Contag

Epithelial tissue malignancies represent 80-90% of all cancer cases making epithelial cells an important target for cancer therapies. Additionally, studies have shown epithelial cells to be targets for cellular reprogramming to pluripotency, an important step for regenerative medicines. Bacterial-based therapies have been explored for treating cancer, and Bacille Calmette-Guerin has become the standard of care for bladder cancer. *B. subtilis* physiology is ideal for a bacteriotherapy chassis organism as it is a generally recognized as safe (GRAS), non-pathogenic, Gram-positive, soil bacterium that respires as a facultative anaerobe. Previous work showed that it can be engineered with transcription factors to function as a modular bacteriotherapy within macrophages. To expand delivery to non-phagocytic cells, *B. subtilis hlyA*, listeriolysin O producing strain, was engineered to express internalin A (InlA), both proteins native to *Listeria monocytogenes.* Internalin A is an adhesin that binds to E-cadherin (E-cad) and uses a "zipper" mechanism for bacterial entry into host cells. E-cadherin is present on the surface of epithelial cells including a mouse mammary carcinoma cell line (4T1), and murine trophoblast stem cells (TSCs). Western blot analysis showed TSCs expressed approximately 2 times more E-cadherin (p=0.019) than 4T1s. Therefore, TSCs were used for InlA-mediated *B. subtilis* delivery to non-phagocytic host cells as an expansion of intracellular bacteriotherapy targets. Cytosolic persistence of the *inlA* strain was confirmed after 2.5 hours of co-incubation. Using a multiplicity of infection (MOI) of 100:1, flow cytometry showed that 6.5% of the TSCs contained bacteria using the *inlA* strain, 20 times more than the 0.3% of TSCs that contained *B. subtilis hlyA* (p<0.0001). *B. subtilis hlyA inlA* was internalized into TSCs at an average of 10-fold more than that of the *B. subtilis hlyA* strain when quantified via a gentamicin protection assay (p<0.01). Potential applications for this chassis strain include bacteriotherapies and cellular reprogramming.

Colon Cancer Screening and Treatment with Engineered Probiotics

Candice Gurbatri, Georgette Radford, Courtney Coker, Jongwon Im, Nicholas Arpaia, Daniel Worthley, Susan Woods, Tal Danino

Bioengineered probiotics enable new opportunities to improve colorectal cancer (CRC) screening, prevention and treatment strategies. Here, we demonstrate selective colonization of CRC tumors with the orally-delivered probiotic *E. coli Nissle 1917* (EcN) in genetically-engineered and orthotopic murine models of CRC, and human CRC patients. We then exploit this neoplasia-homing property of EcN to develop early CRC intervention strategies. To detect lesions, we engineer EcN to produce a small molecule, salicylate, and demonstrate that oral delivery of this strain results in significantly increased levels of salicylate in the urine of adenoma-bearing mice, in comparison to healthy controls. We also assess EcN engineered to locally release immunotherapeutics at the neoplastic site. Oral delivery to mice bearing adenomas, reduced adenoma burden by ~50%, with notable differences in the spatial distribution of T cell populations within diseased and healthy intestinal tissue, suggesting local induction of robust anti-tumor immunity. Together, these results support the use of EcN as an orally-delivered platform to detect disease and treat CRC through its production of screening and therapeutic molecules.

Expanding the Therapeutic Capacity of Engineered Commensal Microbes

Vince Kelly, Yu-Hsuan Yeh, Shannon Sirk

One of the great challenges in the development of biotherapeutics is the high cost associated with these drugs, both in terms of manufacturing cost and the patient's cost for treatment. Dosing regimens requiring repeated administration, trips to a healthcare provider, and large amounts of formulated therapeutic, are a significant driver of these costs. The goal of our research is to leverage the inherent capabilities of the diverse microbial community of the human gut to reduce these costs by acting as therapeutic factories *in situ*. Recent developments in microbial genetics, synthetic biology, and the study of niche-optimized, non-pathogenic members of the human gut microbiota have paved the way for the engineering of commensal bacteria to carry out novel functions. Using species-specific genetic components, therapeutic production can be achieved in both transiently and persistently colonizing members of the gut microbiota. We recently observed that an engineered commensal, *Bacteroides thetaiotaomicron*, can colonize the mouse gastrointestinal tract and deliver a transgenic cargo protein. Tumor- or pathogen-targeting antibody fragments, such as nanobodies, are attractive therapeutic cargo for *in situ* delivery, but are limited in their utility by a lack of secondary effector functions and rapid blood clearance. To address these limitations, we have fused small (15 amino acid) peptides that mimic Fcg receptor- and neonatal Fc receptor (FcRn)-binding epitopes of IgG and albumin to HER2 and *C. difficile* toxin A-targeting nanobodies. These modifications enable native-like receptor engagement and function *in vitro* and *in vivo*. We are also employing rational design strategies to overcome first-pass renal filtration to extend the half-lives of antibody fragments *in vivo*. This work demonstrates the potential utility of engineered antibody fragments either as purified therapeutics or as part of an *in situ* engineered microbial delivery system.

Engineering Protein Secretion of *Bacteroides* Species

Yu-Hsuan Yeh, Suyue Lyu, Vince Kelly, Rahman Rahman Pour, Shannon Sirk

Bacteroides species, one of the most abundant and prevalent bacterial populations in human gut, are capable of long-term, stable colonization of the gastrointestinal tract, making them promising chassis for developing long-term treatment and prevention for chronic diseases. However, the lack of efficient heterologous protein secretion tools prevents their use as engineered, on-site delivery vehicles for protein-based biologic drugs or disease-responsive reporters. Here, we revealed and characterized a group of secretion carriers derived from native secretory proteins of *Bacteroides thetaiotaomicron* that enable the high-titer secretion of functional antibody fragments and reporter proteins in multiple *Bacteroides* species for improving their biomedical applications.

Targeted Receptor-Mediated Programmable Macromolecule Delivery

Xinyi Chen, Wendy Fantl, Lei S. Qi

Gene-editing tools hold great promise for treating many intractable diseases. However, current *in vivo* delivery methods, including viral vectors and nanoparticles, are limited by immunogenicity, low specificity, or short half-life, making them unsafe or ineffective for treatment. Cells are natural 'delivery vehicles' that can produce biomolecule payloads continuously, circulate to targeted tissues, and deliver them through intercellular molecule transfer pathways. However, this potential has not been fully exploited. Trogocytosis is a natural process whereby plasma membrane fragments and their embedded molecules are transferred from a donor cell to a recipient cell[1]. For example, chimeric antigen receptor (CAR) T cells, used in cancer immunotherapy, can acquire antigen molecules from target cells by trogocytosis[2]. Previous studies have shown that trogocytosis is bidirectional[3], indicating that CAR T cells can also be "donor" cells for payload delivery. In this study, we explored using CAR T cell trogocytosis for delivering biomolecules. We found that CAR T cells can transfer membrane-associated molecules to their target cells in an antigen-specific manner within minutes, which can be generalized to different CAR-antigen pairs and target cell types. This process exhibits the hallmarks of trogocytosis, such as dependence on cell-cell contact and actin polymerization, and the detection of the payload on the recipient cell surface. In addition, our preliminary data suggest that the transfer can be potentially separated from T cell cytotoxicity. However, the efficacy of the transferred payloads is limited by their rapid degradation in the recipient cells, which requires further investigation. Our method holds potential to achieve safe, specific, and sustained biomolecule delivery to treat monogenic disorders and other diseases.

1 Joly, E., & Hudrisier, D. (2003). *Nature immunology*, *4*(9), 815-815.

2 Hamieh, M., et al (2019). *Nature*, *568*(7750), 112-116.

3 Joglekar, A. V., Leonard, M. T., ... & Baltimore, D. (2019). *Nature methods*, *16*(2), 191-198.

Site-Specific Brain Therapeutics

Zhimin Huang, Andrei Mitrofan, Shirin Nouraein, Manwal Harb, Jerzy Szablowski

Modulating neuronal activity noninvasively in specific brain regions for extended periods of time would allow for new studies of brain circuitry and provide new therapeutic options with potentially fewer nonspecific off-target effects. Unfortunately, neuromodulation with spatiotemporal precision over long periods of time is a major challenge. Recent studies have shown that spatially precise control of neuronal activity can be achieved with a combination of focused ultrasound blood-brain barrier opening (FUS-BBBO) and delivery of small molecule drugs or viral vectors carrying chemogenetic receptors for gene therapy, in an approach termed Acoustically Targeted Chemogenetics, or ATAC. However, delivery of small molecule therapeutics is short-lived, and their activity is confounded by presence of the opened BBB. On the other hand, chemogenetic approaches such as ATAC allow for long-term activity, but can only be feasibly administered once due to the development of immune response against the vectors, and are hampered by potential risks of toxicity and immunogenicity of the viral vectors. Here, we present a new approach, called Regionally Activated Interstitial Drugs (RAID), that does not require viral vectors, but allows for tunable, noninvasive, long-term neuromodulation with small molecules. RAID relies on noninvasive delivery of an engineered protein enzyme into the brain, which then binds to the brain parenchyma and can locally convert a BBB-permeable inert prodrug into an active neuromodulatory drug. As long as the RAID enzyme is present in the parenchyma, localized neuromodulation can be achieved with systemic administration of the BBB-permeable prodrug even in the absence of BBB opening. Our proof-of-concept studies demonstrated that the engineered RAID enzymes can retain activity in the brain parenchyma for several days. We have also shown site-specific activation of dopaminergic neurons and found that RAID can elicit site-specific behavioral effects. Overall, the RAID paradigm enabled noninvasive, tunable, temporally-resolved, site-specific, non-genetic neuromodulation over multiple days.

Decoding Ligand-Receptor Circuits Involved in Progressive Cardiac Fibrosis

Leili Rohani, Derek Toms, Funda Sar, Sina Safabakhsh, Gurpreet Singhera, Anne Haegert, Yen-Yi Lin, Stephane Lebihan, Lin Wei Tung, Fabio MV Rossi, Colin Collins, Zachary Laksman

Hypertrophic cardiomyopathy (HCM) is a genetic heart disease characterized by progressive fibrosis and risk of sudden cardiac death. Therapeutic options are limited, and no biomarkers are available for monitoring disease progression. We offer a hybrid *cell+gene* therapy platform integrating *in silico* single nuclei RNAseq of patient cardiac tissues with synthetic biology tools (gene regulatory network (GRN) engineering, multiplexing transcription factor (TFs) perturbation, and DNA barcoding) and applied to patient iPSC-derived engineered heart tissues to discover therapeutic targets. Significant differences were revealed between HCM tissues donated to our biobanks by patients with a known sarcomeric pathogenic variant who underwent heart transplantation and healthy control hearts from the International Institute for the Advancement of Medicine.

Single nuclei trajectory analysis showed cell-state transition, specification, and gene signatures suggesting dynamic changes toward fibrotic features in the HCM cardiomyocyte population. Correlation of these features to upstream GRN corroborated the contribution of mutant GRN as a driver of cardiomyocyte fate dysfunction. Laser-focused GRN analysis uncovered novel core TFs responsible for pathways linked to fibrosis, dilated cardiomyopathy, and endothelial-dedifferentiated reprogramming. NicheNet analysis to predict ligand-receptor circuits toward neighboring cells showed aberrant cardiomyocyte-to-fibroblast interaction and communication in HCM. Hypertrophic- (FGF1) and pro-fibrotic- (LAMA2) relevant ligands in diseased cardiomyocytes induced multiple fibrosis-relevant downstream-targeted genes in fibroblast clusters as receivers (receptors). Thus, mutant regulons and aberrant GRNs may initiate the early fibrotic program in cardiomyocytes, resulting in paracrine signaling advancing toward progressive fibrosis. Aberrant cardiomyocyte fate, and ligand-receptor circuits toward niche remodeling in HCM would therefore be primary and secondary effects of mutation.

Our end-goal is to propose and target upstream GRNs in mutant cardiomyocytes as a novel therapeutic target. Our platform could be used for developing disease-modifying therapies by either manipulating upstream mutant cardiomyocyte GRN, or inhibiting released factors regulated by specific regulons (mutant GRNs) using synthetic circuit designs.

Programmable Microbial Communities Via Plasmid Dynamic Control

Hye-In Son, Grayson Hamrick, Lingchong You

Cells can sense and respond to various chemical and physical cues. Synthetic biologists and bioengineers have artificially exploited the gene expression of biological systems to realize innovative applications in several fields[1], such as metabolic engineering, biosensors, living therapeutics, and diagnostics. Despite recent advancements, however, predictable cellular programming is still fundamentally limited by the lack of predictable and versatile biological parts for controlling gene expression levels. Inspired by the robustness of the human adaptive immune system, we introduce a generalizable control system for regulating collective population-level gene expression profiles in synthetic microbial consortia. Combining CRISPR/Cas9-mediated plasmid copy number control, horizontal gene transfer (HGT), and antibiotic-mediated selection, we demonstrate that our platform can amplify gene expression dynamic range by several orders of magnitudes in addition to the typical transcription factor-mediated gene regulation methods. In addition, we show that our system can adopt a user-defined temporal program to dynamically regulate the population-level gene profile and orthogonally regulate multiple genes. Applying our design principle as a proof of concept, we constructed a biosensor that can detect tetrathionate, a biomarker for inflammatory bowel disease, with a wide dynamic range.

1 Khalil, A. S. & Collins, J. J. Synthetic biology: applications come of age. *Nat Rev Genet* **11**, 367-379, doi:10.1038/nrg2775 (2010).

Adaptive Evolution of *E. coli* for Accelerated Cell Growth Rate

Hye-In Son, Grayson Hamrick, Cesar Villalobos, Kyeri Kim, Lingchong You

The long incubation times required for dense cell growth are a major rate limiting step for many types of microbiology experiments. *Escherichia coli* has served as a popular model organism in various aspects of modern biotechnology, biological engineering, and molecular biology due to its well-understood genomic background, ease of culture, and *relative* growth rate. However, many experiments still require hours or days to complete due to the speed limit imposed by natural selection on microbial growth rates. In this work, we engineer a synthetic gene circuit to artificially select for faster-growing subpopulations of *E. coli*. Our platform implements CRISPR/Cas9 to impose a growth rate threshold for survival. We demonstrate that *E. coli* cells carrying our circuit can survive only if they grow fast enough. Using this CRISPR/Cas9-integrated circuit, we accelerated the growth rate of E. coli by 70% in 28 days in a long-term adaptive evolutionary experiment. This work introduces a novel strategy for engineering a bacterial host for accelerated biotechnology and molecular biology research.

Addressable and Adaptable Intercellular Communication Via DNA Messaging

John P. Marken, Richard M. Murray

Engineered consortia are a major research focus for synthetic biologists because they can implement sophisticated behaviors inaccessible to single-strain systems. However, this functional capacity is constrained by their constituent strains' ability to engage in complex communication. DNA messaging, by enabling information-rich channel-decoupled communication, is a promising candidate architecture for implementing complex communication. But its major advantage, its messages' dynamic mutability, is still unexplored. We develop a framework for addressable and adaptable DNA messaging that leverages all three of these advantages and implement it using plasmid conjugation in *E. coli.* Our system can bias the transfer of messages to targeted receiver strains by 100- to 1000-fold, and their recipient lists can be dynamically updated *in situ* to control the flow of information through the population. This work lays the foundation for future developments that further utilize the unique advantages of DNA messaging to engineer previously-inaccessible levels of complexity into biological systems.

Modeling Gut-Skin Interaction in the Immunopathogenesis of Psoriasis

Gowri Yathishchandran Subedar, Michaelle Mayalu

The gut-skin axis is a bi-directional communication between the gut and the skin. Any perturbation to this axis can lead to skin, gastrointestinal comorbidities. Psoriasis is a common skin disorder in which this axis is perturbed. Psoriasis patients display variation in the skin and gut microbiota composition and abundance that leads to plasma concentration variation of glucose, lipid, amino acid and nucleotide metabolites [Chin. Med. J. 133 (2020), pp. 2976-2986]. Further, the neurotransmitters released by certain bacteria modulate the immune response [Dermatol. Ther. 11 (2021), pp. 25-38]. Genetic factors, like HLA-Cw6 allele variability, modify proteins, initiating a positive feedback loop of cell proliferation and sustained immune response [Dermatol. Online J. 235 (2019), pp. 91-100]. Quantitative studies of psoriasis previously model only the keratinocyte-immune cell interactions, neglecting the effect of the interactions with gut and skin [Sci. Rep. 11 (2021), pp. 1-13]. We show that these interactions can constrain the system and lead to system bistability, which suggests the presence of healthy and diseased states. We further describe the effect of bacterial interactions, initial cell concentrations and cytokine effects on the transition between the bistable states. Our results demonstrate the importance of modeling the 'gut-skin' axis in the context of the immune response in psoriasis. We anticipate our study to be a starting point for more sophisticated mathematical models of psoriasis, for example, by obtaining representative values for the model parameters using data-driven methods from patient data. Furthermore, numerical simulations carried out simultaneously with clinical trials can provide further insights on the pathogenesis of psoriasis and identify potential therapeutic targets.

A Deconstructed Geminiviral Vector Approach to Designer Protein Production in *N. Benthamiana*

Al Watson, Kevin Morey, Christie Peebles

Recent advances in synthetic biology have allowed for the development of plants as biofactories to produce therapeutic pharmaceutical proteins such as monoclonal antibodies, viral vaccines, and enzymes for replacement therapy. Scalable pharmaceutical protein expression protocols minimize time and cost by taking advantage of the previously described natural transient gene expression systems of *Agrobacterium tumefaciens*. However, the typical methods of expressing heterologous proteins through agroinfiltration coupled with a strong constitutive promoter often do not result in a protein titer high enough for certain experimental and industrial applications. Therefore, our lab utilized a synthetic biology approach that combined a strong constitutive promoter with a viral "cassette" vector in order to increase the protein titer. Here we will describe the biotechnology for overexpression of the heterologous protein serglycin in *Nicotiana benthamiana* leaves, which involves: 1) construction of a geminiviral-based expression cassette vector, 2) scalable agroinfiltration-based delivery into plant leaves, 3) subcellular targeting to ensure proper packaging of the heterologous protein. Attaching a polyhistidine-tag to the protein allowed for simple isolation and purification of the protein of interest. The synthetic biology approach we describe ensures overexpression of the heterologous protein serglycin in plants.

Signal Crosstalk Sets Activity States within Diverse Microbial Communities

James Boedicker

Bacteria communicate to regulate macro-scale patterns of gene expression and to coordinate group behaviors. Cells secrete, sense, and respond to small chemical signals in a process known as quorum sensing. Although the process of quorum sensing is well understood, most studies focus on signaling within well-mixed, monocultures of bacteria. In many contexts, microbial communication is impacted by both biological and spatial complexity. Within diverse bacterial communities, signal exchange is modulated by interactions between different cell types. For example, two different cell types can produce chemically similar signal molecules. Similar signals can interact with the same receptor proteins, leading to signal crosstalk. Crosstalk influences the expression of signal-regulated genes. Recently, we characterized crosstalk between five naturally occurring strains of *Bacillus subtilis* by measuring the response of each strain to mixtures of signal. Experimental measurements of crosstalk were used to construct a neural network/perceptron model, and the model accurately predicted how bacteria interpret complex mixtures of signals in the environment. We also explored how communication within such cellular networks is impacted by spatial structure, revealing that signaling interactions can lead to a percolation transition of spatial patterns of activity. Our work combines experimental and theoretical approaches to understand how community composition, signal crosstalk, and spatial structure influence quorum sensing within bacterial communities.

Kluyveromyces Marxianus – a Green Cell Factory for Organic Acid Production

Mackenzie Thornbury, Iain Summerby-Murray, James Dhaliwal, Jaya Joshi, Lauren Narcross, Adrien Knoops, Vincent J.J. Martin

Organic acids, such as fumaric acid, feature heavily in manufacturing due to their ease of polymerization into diverse industrially relevant compounds. Currently, many of these organic acids are derived from chemical conversion of fossil fuels in energy intensive and greenhouse gas-producing processes. To mitigate these environmental concerns, our group is engineering the yeast *Kluyveromyces marxianus* to act as a platform for organic acid biosynthesis from renewable feedstocks. *K. marxianus* is an attractive host because it is the fastest growing eukaryote, able to utilize a broad range of substrates, thermotolerant to 50 °C, and acid-tolerant to pH 2.5. Leveraging these characteristics, we aim to valorize the lactose present in milk permeate, a milk industry waste-product, by converting it into the dicarboxylic organic acid fumarate.

Genetic tools available for manipulating *K. marxianus* are limited. We are expanding this toolkit using CRISPR activation, inhibition, and deletion (CRISPR-AID) tools aimed at creating a genome-wide CRISPR library. These tools can be used to induce fumarate production through rational design and high-throughput screening. We are currently building a CRISPR-AID genome-wide guide library that will enable screening for higher fumarate production and improved organic acid tolerance. Thus far, we have successfully tested the CRISPR-AID system to show all three CRISPR-directed perturbations in *K. marxianus*. We have also used the CRISPR deletion tools to increase fumarate production from 0 to 260 mg/L by deleting the endogenous fumarase, and thus preventing fumarate from re-entering the TCA cycle. Further work is underway to express a highly active fumaric acid synthesis pathway. Collectively, these techniques can result in a new platform for organic acid production from dairy by-products while also providing new genetic tools to unlock the metabolic engineering potential of *K. marxianus*.

Engineering Novel Microbes for Upcycling Waste Plastic and Solving Climate Crisis

Jinjin Diao, Yuxin Tian, Yifeng Hu, Tae Seok Moon

The use of petroleum-based products has negative impacts on our planet, leading to climate crisis. Polyethylene terephthalate (PET) represents 12% of global solid waste. PET chemical recycling has been an option to solve this global problem, but it suffers from its relatively high process cost and the extremely low price of virgin PET. One solution to address this issue is to upcycle waste PET rather than recycle it to generate the same PET typically with low quality. PET upcycling can be achieved by depolymerizing PET into terephthalic acid (TPA) and ethylene glycol (EG) and biologically converting these monomers into value-added products. However, there are only a handful of reports demonstrating microbes capable of growing on both TPA and EG generated from PET as sole carbon sources. To overcome this limitation, we have performed strain screening to discover a *Rhodococcus* strain RPET that can grow well on the alkaline hydrolysis products of PET as the sole carbon source without any purification step. Notably, this strain can grow on a mixture of TPA and EG at extremely high concentrations and high osmolarity resulting from alkaline hydrolysis and pH neutralization. The resultant media supported RPET's growth without any purification and sterilization step except for their dilution. Many synthetic biology tools, developed for a related species *Rhodococcus opacus* (1,2), were functional in RPET, facilitating its engineering. We will discuss our effort to develop this novel chassis for waste PET valorization with PET conversion into carotenoids, fatty acids, and muconate (3,4). In addition, we will discuss our efforts to valorize mixed plastic waste by converting the mixture into these products.

1. DeLorenzo, Rottinghaus, Henson, Moon. 2018. ACS Synth. Biol. 7:727-38
2. DeLorenzo, Diao, Carr, Hu, Moon. 2021. ACS Synth. Biol. 10:786-98
3. Moon et al. PCT Patent
4. Diao, Hu, Tian, Carr, Moon. 2023. Cell Rep. 42, 111908

Scalable, Equipment-Free Affinity Protein Purification By Genetically Engineered Buoyant Nanoparticles

Zongru Li, George Lu

Affinity chromatography is widely used in laboratory settings due to its versatility in producing high-purity products from a variety of targets. However, scaling up the method to deal with thousands of liters of liquid and above at the industrial level is facing considerable challenges due to the high cost of affinity resins and chromatography columns. To address this challenge, we have demonstrated that gas vesicles, a class of protein nanostructures evolved in photosynthetic microbes, can be genetically engineered with high-density affinity tags on their surfaces. This modification enables efficient affinity binding of the protein of interest, followed by buoyancy-driven isolation. Our method has been successfully demonstrated with the purification of fluorescent proteins, and the yield and purity have been compared to that of standard nickel affinity chromatography. Uniquely, these buoyant nanoparticles can be processed through centrifugation or flow-based methods, which can be readily scaled up in industrial settings. Moreover, to entirely eliminate the need for equipment, we have demonstrated that controlling the aggregation of our nanoparticles allows flotation over multiple centimeters in an hour under normal gravity. This development sets the stage for the use of our technology in biomanufacturing and diagnostics under low-resource settings. Our future directions will focus on designing the process and workflow of the method to efficiently purify industrial enzymes and protein biologics. Overall, our research provides a novel platform for industrial-scale purification of recombinant proteins that reduces reliance on expensive affinity resins and columns. Furthermore, the high-density tagged surface of gas vesicles offers potential use as an assembly platform for a variety of nanotechnologies and biomaterial applications.

A Toolkit for Engineering *Geobacillus* for Degradation of Waste Produced By the Built Environment

Matt Rogan, Paul James

The Hub for Biotechnology in the Built Environment aims to produce living and responsive buildings that can metabolize their waste to reduce pollution, produce energy and other high valued products. Bacteria with interesting phenotypes can be used as microbial cell factories for production of high valued products from waste. One candidate that can act as a microbial cell factory for the built environment is the Gram-positive bacterium *Geobacillus*, for its broad substrate utilization, industrially important enzymes and its ability to grow at high temperatures.

To better utilize this genus, a toolbox of basic genetic parts (constitutive promoters, ribosome binding sites, insulators, terminators, and signal peptides) was made for engineering the *Geobacillus* genus. To improve predictability of genes expression, the genetic insulator RiboJ and synthetic ribosome binding sites of known translation initiation rates were used to standardize expression of different coding sequences, allowing normalized expression of GFP and mOrange when using the same promoter. To allow secretion of heterologous proteins, signal peptides were identified using proteomics, bioinformatics and characterised using mOrange to confirm secretion. Finally, rho-independent terminators with varying terminator efficiencies were characterised.

Using these parts, genes involved in alkane degradation were expressed to expand the existent alkane utilization capabilities in a selection of *Geobacillus* species. The aim for this expansion was to utilize a waxy material rich in hydrocarbons, produced from cold plasma assisted pyrolysis of non-recyclable plastics. Using this toolkit, D-glucaric acid (a greener alternative to phosphate-based detergents) will be produced using *Ino1*, *MioX* and *udh*. Additionally, a statistical design of experiments approach will be used to optimize production by exploring the effects of gene expression and growth conditions on D-glucaric acid production. Through pairing thermochemical breakdown of non-recyclable plastics with industrial production in an engineered microbial cell factory, a circular economy within the built environment can be achieved.

Synthetic Microbe-to-Plant Communication Channels

Alice Boo, Tyler Toth, Qiguo Yu, Alexander Pfotenhauer, Scott Lenaghan, Neal Stewart, Christopher A. Voigt

Interactions between plants and microbes play a crucial role in pest control, nutrient scavenging, and adaptation to environmental change. The complex communication between thousands of microbial species and plants is mediated by a vast chemical language and complex regulatory networks. To facilitate the design of plant-microbial communities, we have developed modular interkingdom communication channels using a root-associated bacterium, *Pseudomonas putida*, that detects environmental stimuli and transmits this information to the plant using the quorum sensing molecule p-coumaroyl-homoserine lactone (pC-HSL). We validated that the plant pC-HSL receiver in *Arabidopsis thaliana*and *Solanum tuberosum* successfully detects the presence of pC-HSL produced by the *P. putida* sender strain in hydroponic and soil systems. We demonstrated the modularity of this system by swapping bacteria to make the same plant respond to different chemical stimuli. We also showed the capacity of bacteria to integrate multiple stimuli and relay a single output to the plant by creating simple logic gates in *P. putida*. This programmable communication between engineered bacteria and plants will enable microbial sentinels to transmit information to crops and provide the building blocks for designing artificial consortia with distributed functions.

A Sentinel Plant Platform for Visualizing the Rhizosphere Dynamics

Yunqing Wang, Gözde Demirer

The rhizosphere, which is the critical boundary between plant roots and soil microbes, is an essential component of soil health and sustainability in agriculture. Microbes in the rhizosphere contribute substantially to climate change by emitting greenhouse gases. However, we do not have real time technologies to visualize and quantify emissions from agricultural soils. Here, we establish a novel sentinel plant platform for monitoring the rhizosphere dynamics, specifically emissions of nitrous oxide (N_2O), in which below-ground microbial production of N_2O-generating enzymes will be converted to above-ground visible signals in plant leaves that can be detected in real time at large scale using drones or satellites.

To enable the communication between soil microbes and the sentinel plant, we are leveraging the *p*-Coumaroyl-homoserine lactone (*p*C-HSL) as the trans-kingdom orthogonal signal, which is found only in three bacteria species[1]. This sentinel plant should be able to interface with *p*C-HSL signals, whose secretion is tightly regulated by the expression levels of target genes across rhizosphere microbes, via root-to-shoot uptake and activation of aboveground reporter circuits for visual readout. Our results to date have shown that plant roots can uptake *p*C-HSL and transport it to the leaves. We are currently engineering sensitive, specific, and reversible reporter circuits to visually display *p*C-HSL activation in leaves with a wide dynamic range, which is based on a transcriptional activator and fluorescent or luminescent reporter driven by a synthetic promoter consisting of multicopy of the cis-regulatory sequences upstream of the minimal 35S promoter.

The sentinel plant platform enables the minimally invasive, continuous monitoring of the expression levels of target genes across rhizosphere microbes with an aboveground readout, providing a deeper understanding of the rhizosphere and its impact on sustainability-relevant processes.

(1) Schaefer, Amy L., et al. 'A new class of homoserine lactone quorum-sensing signals.' Nature 454.7204 (2008): 595-599.

Genetically-Encoded Delivery Vehicles for Applications in Plant Synthetic Biology

Mark Legendre, Gözde Demirer

The powerful CRISPR-Cas gene editing system has transformed our synthetic biology capabilities, making possible highly-specific and efficient genetic modifications. However, the benefits of CRISPR have failed to permeate the plant synthetic biology community as effectively as they have its bacterial and mammalian counterparts. This is largely imparted by the unique physical barriers to CRISPR reagent delivery encountered in plant systems, including a rigid cell wall that hinders intracellular transport of extracellular biomacromolecules. Nanotechnology has emerged as a promising solution to this inherent transport barrier, enabling the delivery of silencing RNAs and expression plasmids via nanocarriers such as carbon nanotubes and gold nanorods, but these delivery platforms are limited in cargo diversity, scalability of synthesis, and translatability to agricultural fields.

To overcome these limitations and enable facile cargo delivery to plants for various synthetic biology applications, a Genetically-encoded Delivery Vehicle (GDV) was adapted from the Tobacco Mosaic Virus (TMV) viral capsid. Using a highly efficient translational system, TMV-GDVs were synthesized *in planta* and isolated for downstream modification and application in biomacromolecule delivery. Bioorthogonal click chemistry was leveraged for peptide functionalization of the TMV-GDV surface, thus enabling various functionalities. Most notably, *in planta* fluorescence characterization demonstrates successful particle entry into the epithelial cells of tobacco leaves and offers a method for quantifying modulations in particle delivery efficiency. Additionally, a cell-penetrating peptide (CPP) surface treatment was explored as a method for improving delivery efficiency and for electrostatically grafting negatively-charged plasmid DNA to the GDV surface. GDVs offer a biodegradable, genetically-encoded, and scalable alternative to more traditional nanocarriers used in plant synthetic biology, boasting a two-pronged modulation strategy at the material level that exploits both protein engineering and chemical surface modification to realize a new and powerful tool in the plant synthetic biology toolkit.

Leveraging Aromatic Amine Bioproduction in Stress-Resistant *Pseudomonas Putida* with Crispra/I Gene Regulatory Tools

Cholpisit Kiattisewee, Ian Faulkner, Jesse Zalatan, James Carothers

Pseudomonas putida is an attractive chassis for industrial bioproduction due to its tolerance to high concentrations of valuable products and resistance to harsh bioprocessing conditions. The recent creation of tools for multi-gene CRISPR gene regulation is accelerating the development of *P. putida* as a platform for bioproduction. We have shown that CRISPR gene activation (CRISPRa) programs operating through the expression of 3 guide RNAs can be engineered to produce *p*-aminocinnamic acid (p-ACA). *p*-ACA is a precursor for high-performance polymer synthesis whose toxicity prevents it from being efficiently produced in model microbes like *E. coli*. By fine tuning enzyme expression levels in this system, 100 mg/L *p*-ACA titers could be produced directly from glucose. We have seen that similar production titers can be achieved even when as many as 6 guide RNAs are expressed in the system. Taken together, these results suggest that larger multi-guide CRISPR programs that could regulate expression from multiple heterologous and endogenous targets and further increase flux toward valuable products are within reach. We will present these results and outline a strategy to use multi-gene CRISPRa/i programs to accelerate metabolic engineering in non-model bacteria as platforms for chemical production.

Carbon-Negative Production of Fuels and Chemicals from Industrial Waste Gases Using Engineered Microbes

Fungmin Eric Liew, Robert Nogle, Rebecca Ostien, Shivani Garg, Ching Leang, Steve D. Brown, Sean D. Simpson, Michael Koepke

There is increasing interest in biomanufacturing to utilize feedstock that does not compete with food production and has an overall carbon-negative footprint. By tapping into industrial waste gas (e.g., steel mill plant, biorefinery) and synthesis gas generated from agricultural residue and non-recyclable municipal solid waste, gas fermentation is a bioprocess that converts carbon oxide gas streams into carbon-negative fuels and chemicals. Currently LanzaTech successfully started up three commercial scale gas fermentation plants that produce ethanol (45-60k MTA per plant) and biomass (for animal feed). In addition to drop-in fuel applications, ethanol can be chemically converted to jet fuels, fabric and packaging materials.

By employing a collaborative and multi-disciplinary approach to engineer these gas-utilizing microbes, we drastically expanded the product profile of these microorganisms to produce more than 50 diverse chemicals, ranging from short chain carboxylic acids and alcohols to aromatics and terpenes (1). Two such examples are the production of non-native acetone and isopropanol at high productivity and selectivity from industrial waste gas at pilot scale (2). To accelerate strain engineering effort, cell-free prototyping was utilized to test enzyme variants and expression levels, which resulted in gas-utilizing strains with improved 1-hexanol production (3). In an effort to compartmentalize heterologous pathways of interest to alleviate toxicity and minimize off-target biochemical reactions, we have successfully demonstrated encapsulation of reporter proteins into microcompartments in gas-utilizing microbes.

References

1. Köpke, M. and Simpson, S. (2020). Pollution to products: recycling of 'above ground' carbon by gas fermentation. Current opinion in Biotechnology 65: 180-189.
2. Liew, F.E. *et al.* (2022). Carbon-negative production of acetone and isopropanol by gas fermentation at industrial pilot scale. Nature Biotechnology 40: 335-344.
3. Vögeli *et al.* (2022). Cell-free prototyping enables implementation of optimized reverse β-oxidation pathways in heterotrophic and autotrophic bacteria. Nature communications 13: 3058.

Reproducible Bioluminescence Sensors for Real Time Biological Nitrate Sensing: Biofilm's Influence on Reporter Expression

Mary Machado

Whole-cell bio-reporters are promising candidates for low-cost environmental sensing of both pollutants and bioavailable nutrients. Previous versions of these sensors have been able to detect bioavailable nutrients such as nitrates at the low levels found within environmental systems. Yet difficulties remain in creating biosensors that are suitable for long term deployment in the field. Achieving good specificity, sensitivity and robustness in these sensors has proved especially challenging. This work describes the creation of a model to improve precision within whole cell biosensors. During this work, biofilm formation and luxAB expression were assessed in a self-bioluminescent strain of cyanobacteria exposed to the environmental conditions of the marine environment. This work used methods including QCM-R, luminometry and fluorescence spectroscopy to identify variations in lux expression during biofilm formation and to create a training set for an in-depth model of sensor reproducibility across low cost sensors. Sensors in this work consisted of low cost housing which included a cell cartridge, a light to frequency sensor and a microprocessor. This work found that low cost sensing equipment was both sensitive and specific and could be an ideal candidate for continuous monitoring of marine and estuarine systems.

Screening of Millions of Colonies Isolated in Hollow Microparticles Based on Natural Product Metabolite Production.

Cayden Williamson, Moriel Dror, Mark van Zee, Joshua Misa, Yi Tang, Dino Di Carlo

Engineering microbial hosts to express plant natural product (NP) pathways is a sustainable and scalable method of producing high-value compounds, including food additives and pharmaceuticals. While *Saccharomyces cerevisiae* is a suitable host for expression of plant alkaloid enzymes and production of the target NP, generating these NPs at sufficiently high titer for industrial application remains challenging due to enzymatic bottlenecks unique to the metabolic environment of the native host. Advances in synthetic biology have given researchers powerful tools to generate massive bio-enzymatic mutant or genome-editing libraries that can be screened for production-enhancing targets using a detectable small molecule biosensor that correlates with target production. However, optimization of secondary product output is hindered by lack of technology to isolate strains with distinguished mutations at a rate comparable with strain library generation.

We offer the PicoShell cell evolution platform as a method to directly interrogate intracellular metabolite production at several orders of magnitude higher rates than current colony picking techniques – hundreds of thousands of clones per hour. PicoShells are hollow microparticles with porous shells that function as picolitre-scale culture chambers for proliferation of monoclonal microbial colonies. The output of standard *in vivo* fluorescent probes or pathway-related fluorescent metabolites is often not detectable at single-cell levels. Single-cell expansion into hundred-cell colonies within PicoShells allows clones to amplify their fluorescent signal by orders of magnitude, enabling clones to be selected through fluorescence activated cell sorting. In a single sort workflow, we performed multiplexed colony selection using growth and metabolite production phenotypes as selection probes. We selected subpopulations of differently producing yeast colonies, released and recultured them, resulting in populations enriched for a desired set of phenotypic traits. Through repeated selection of high-producing colonies, PicoShells can be used for accelerated "continuous evolution" to enhance target metabolite yields for the purpose of sustainable microbial biomanufacturing.

One Material to Bind Them All: Bottom-up Engineering of Living Materials for Selective Rare Earth Element Binding

Swetha Sridhar, Caroline M. Ajo-Franklin

Rare Earth Elements (REEs), such as lanthanides, are critical for defense, energy, and medical applications. However, despite their importance, these elements require harsh chemical extraction methods that are expensive, laborious, and environmentally toxic. There is a pressing need for scalable, sustainable, and facile extraction and enrichment methods that can purify REEs from synthetic and natural waste streams, such as industrial effluents and acid mine drainage.

To extract and enrich REEs, we leverage our bottom-up engineered living materials as platforms for a protein, lanmodulin, that binds to lanthanides with picomolar affinity. Our living material scaffolds enable high-density surface display of lanmodulin, thereby maximizing selective REE binding. We also analyze the binding site of lanmodulin to enhance selectivity towards different REEs. Through our engineered living materials, we can enrich REEs from waste streams more scalably, affordably, and sustainably.

Investigating the Role of Amino Acid Residues in μ-Opioid Receptor Function Using Site Saturation Mutagenesis in a Yeast Biosensor

Manpreet Vilkhu, Christien Dykstra, Bjorn Bean, Malcolm Whiteway, Vincent J.J. Martin

Opioids are potent pain relievers that bind to specific G protein-coupled receptors (GPCRs) in the brain and other areas of the body, inducing an intracellular response that results in analgesia. While the World Health Organization recognizes opioids as essential pain medication, they can also be highly addictive and cause severe side effects. Understanding the function of each amino acid within a receptor may help explain opioid addiction and lead to the development of opioid-like drugs with similar analgesic properties but without addiction risk. Despite the identification of the structure of the human μ-opioid receptor, the role of each amino acid residue in receptor function remains unknown.

Scanning site saturation mutagenesis (SSSM) can aid in understanding the functional components of μ-opioid receptor residues. SSSM involves systematically mutating each amino acid residue in the receptor and screening the variants for receptor function. However, SSSM requires high-throughput screening, which can be challenging in mammalian cells but is possible in yeast. Our laboratory has recently developed a humanized yeast-based opioid biosensor by functionally connecting opioid receptors to an endogenous signaling system resulting in increased expression of GFP in the presence of opioid agonists. However, this biosensor has low sensitivity and selectivity.

We have designed an SSSM library targeting predicted structurally and functionally relevant regions to test in our yeast biosensor. The library variants will be screened on various opioid analogs and sorted by fluorescent intensity. The desired variants will be sequenced using next generation sequencing to reveal the unique roles played by each amino acid in binding and activation. This study will contribute to the development of a reliable cell-based pharmacological screening platform with improved sensitivity and selectivity for quantifying opioid production, discovering novel ligands from pharmacological libraries, and enhancing the overall understanding of the structure/function relationship of GPCRs.

Macroscopic Engineered Living Materials from Bacteria

Sara Molinari

Naturally occurring biomaterials grow bottom-up from cells into macroscale structures. Engineered living materials (ELMs) are composites of living cells incorporated into a biopolymer matrix. They are inspired by natural biomaterials but use synthetic biology to introduce tailored properties. While cells confer functionality to ELMs, the matrix defines their mechanical and physical properties. For this reason, the ability to engineer the collective self-organization of cells through a genetically encoded synthetic matrix has been a longstanding challenge in the ELM field. Previous to my work, most macroscopic ELMs either required significant processing to be assembled or were scarcely tunable. I addressed this limitation, by creating the first macroscopic de novo ELM, which grows from genetically-engineered bacteria. To achieve this goal, I engineered *Caulobacter crescentus* to display self-interacting proteins by replacing the central crystallization domain of the native surface layer monomer (RsaA) with a structural domain made of elastin-like polypeptides. In this way, the hydrophobic C-termini of the RsaA become exposed and able to self-interact. I thus created a synthetic extracellular matrix that mediates cell-cell interactions. This protein was also found to be partially secreted, forming a matrix able to hierarchically organize cells over four orders of magnitude, resulting in the growth of centimeter-scale living materials. The mechanical properties of de novo ELMs can be tuned genetically over a 25-fold range by altering the structural domain of the matrix protein. I showed that these materials can be also functionalized from complex enzyme mixtures allowing them to perform biological catalysis. Moreover, they can be employed to filter cadmium from contaminated solutions. De novo ELMs can be reshaped and used as cementing agents, they can be desiccated and reseeded to form new material. This study lays the foundation for growing ELMs with defined physical and mechanical properties, paving the way toward growing multifunctional, self-regenerating materials.

Fibre Fusion: Circular Manufacturing of Water Repelling Bacterial Cellulose through a Biological Approach

Paul James, Katie Gilmour, Christopher Markwell, Mahab Aljannat, Yunhong Jiang, Jane Scott, Martyn Dade-Robertson, Meng Zhang

There is increasing interest in utilising synthetic biology approaches to produce engineered living materials. In nature, biological materials are synthesised and functionalised by constructing complex structures composed of different molecules. For example, biofilms, which provide protection for bacterial cells, constitute a range of functionalities through the addition of extracellular molecules including proteins which enhance strength, flexibility, and hydrophobicity.

We explored the possibility of producing functionalised bacterial cellulose, inspired by biofilm formation. By creating new sustainable methods to produce modified bacterial cellulose, it may have broader applications. BslA is a hydrophobic protein involved in biofilm formation of *Bacillus subtilis*, which when applied to bacterial cellulose, could produce a versatile, water repellent biomaterial that could then be used in textile applications such as a sustainable leather-like textile.

BslA was expressed recombinantly with a cellulose-binding module and then applied to bacterial cellulose (ex-situ) or added to initial growth media (in-situ) of cellulose-producing bacteria, *Komagataeibacter xylinus*. The topography of the resulting cellulose was observed using atomic force microscopy and changes to hydrophobicity were analysed using water contact angle testing. Whilst BslA was able to increase the hydrophobicity of the cellulose surface, it was found that the orientation of the cellulose-binding module may interfere with this functionality and further alterations to the protein sequence of BslA and position of the cellulose-binding module may promote a more significant increase in hydrophobicity. Testing showed an increase in the fire retardancy of BslA treated fibers and results from mechanical testing of the treated cellulose supported the conclusion that BslA with cellulose binding domains alters the formation of cellulose pellicles, resulting in an increase in tensile strength and decrease in Young's modulus. Therefore, it was concluded that functionalising cellulose using a recombinant protein did increase the hydrophobicity to the same extent as conventional chemical treatments.

Dynamic Gene Expression in Engineered Living Materials for Bioproduction

Widianti Sugianto, Benjamin Tickman, Cholpisit Kiattisewee, Alshakim Nelson, James Carothers

Engineered living materials (ELMs), comprised of engineered microbes in biomaterial scaffolds, have been reported to sustain long-term metabolic activity useful for on-demand bioproduction. However, the extent to which genetic activity can be sustained in ELMs has not been fully explored. In this work, we investigate the ability of microbial ELMs to sustain gene expression when continuously cultured for multiple weeks. To do this, we apply inducible CRISPR-based expression tools on *E. coli* encapsulated in F127-bisurethane methacrylate (F127-BUM) hydrogels and develop a plate reader-based assay to directly measure fluorescent protein expression in these ELMs over time. We find that gene expression in ELMs is sustained for multiple weeks, where ELMs extend the inducible response in bacteria by 2.5 weeks longer than that in planktonic, liquid cultures. This study also reveals that ELMs maintain responsiveness to multiple induction cycles, retaining up to 97% of the initial gene expression capacity. We demonstrate that bioproduction in ELMs can be repeatedly induced (cycled) for multiple weeks. We implement inducible CRISPR transcriptional activation (CRISPRa) programs that regulate bioproduction of pteridine, a compound used for phenylketonuria (PKU) treatment. We find a resemblance between the inducible dynamics of the bioproduction and the fluorescent protein expression. This suggests that ELMs are genetically and metabolically active in long-term continuous culture and can be dynamically controlled via CRISPR genetic programs. Taken together, these results show that ELMs retain long-term, programmable expression capacity that can be used for sustained bioproduction, foreshadowing the creation of next-generation ELMs for controllable, scalable, and versatile on-demand bioproduction.

Thermoresponsive Living Therapeutic Materials - Darobactin-Releasing Bacterial Hydrogels for Treating Chronic Pathogenic Infections

Sourik Dey

Targeted drug delivery through engineered microorganisms is a rapidly growing approach due to the extensive versatility of synthetic biology tools. Several beneficial bacteria are currently being used as a safe microbial chassis for developing sustainable drug production platforms. Our focus is to engineer an endotoxin free strain of E.coli to release a potent antimicrobial, Darobactin, to selectively eliminate Gram-negative pathogens. Darobactin production is regulated by an externally applied and sharply controlled temperature trigger for the sustainable use of engineered living therapeutics at diverse sites in the host anatomy. Peripherally induced heating can activate and maintain the drug production above the normal physiological temperature of 37oC with a high order fold change. The engineered bacteria will be securely encapsulated in biocompatible hydrogels to protect the bacteria from harsh conditions in the body, while preventing their escape to the surrounding tissues. The long-term sustenance and temperature induced drug release of the bacterial populations in hydrogels can act as an advanced therapeutic strategy to complement the existing treatment methodologies and provide better resolutions in eliminating chronic pathogenic infections

Secretion-Catalyzed Assembly of Protein Biomaterials on a Bacterial Membrane Surface

Qi Xie, Stephen D. Fried

Protein-based biomaterials have played a key role in tissue engineering, and additional exciting applications as self-healing materials and sustainable polymers are emerging. Over the past few decades, recombinant expression and production of various fibrous proteins from microbes have been demonstrated; however, the resulting proteins typically must then be purified and processed by humans to form usable fibers and materials. Here, we show that the Gram-positive bacterium *Bacillus subtilis* can be programmed to secrete silk through its translocon via an orthogonal signal peptide/peptidase pair. Surprisingly, we discover that this translocation mechanism drives the silk proteins to assemble into fibers spontaneously on the cell surface, in a process we call secretion-catalyzed assembly (SCA). Secreted silk fibers form self-healing hydrogels with minimal processing. Alternatively, the fibers retained on the membrane provide a facile route to create engineered living materials from *Bacillus* cells. This work provides a blueprint to achieve autonomous assembly of protein biomaterials in useful morphologies directly from microbial factories.

Reprogramming Chromosome-Free Synthetic Cells As a Platform for Targeted Disease Therapy

Yutong Yin

SimCells (simple cells) are chromosome-free bacteria controlled by designed gene circuits, which can bypass the interference of the native gene network in bacteria and eliminate the risk of uncontrolled bacterial growth. We reprogrammed the SimCells and mini-SimCells to serve as "artificial immune cells" that specifically target antigen and locally release drugs in a highly controlled manner for disease therapeutics. We engineer SimCells to display nanobodies on the surface for targeting antigens, such as carcinoembryonic antigen (CEA), an important biomarker found commonly in colorectal cancer cells. We further enhanced the cancer-killing effect by incorporating an aspirin/salicylate inducible gene circuit that converts salicylate into catechol, a potent anticancer drug. This work highlights the potential of SimCells and mini-SimCells as reprogrammable platforms for therapeutics and lays the foundation for applying synthetic biology to medicine.

Synthetic Biology-Enabled Wearable Biosensors for Health Monitoring

Behide Saltepe, Maggie Hicks, Xinyi Wan, Baojun Wang, Chris French

Advances in synthetic biology tools have paved the way for engineering of various types of biosensors for health monitoring. Continuous monitoring of biomarkers provides cumulative information about health status and deviations over time. Although the gold standard for health monitoring is blood, sweat also contains many of the same biomarkers and can be passively sampled enabling an alternative non-invasive method of health monitoring. Recent developments in wearable device technology make frequent and non-invasive monitoring possible, increasing individual compliance. Wearable biosensors would allow multiplex detection of biomarkers to monitor physiological parameters related to disease, stress, dehydration and exhaustion which are associated with the biomarker concentration. In this project, we investigate possible formats for biosensors which can be turned into wearable devices as adhesive patches allowing non-invasive health and fitness monitoring from sweat. Here we have tested four different types of biosensors based on whole-cell biosensors, cell-free transcription/translation-based biosensors, cell free transcription-only biosensors, and engineered protein or aptamer-based biosensors. In all cases, these biosensors are designed to provide a fluorescent output which can be detected and quantified using simple apparatus such as a mobile phone camera with suitable filters. After characterizing biosensors in multi-well plates, we have tested them via hydrogel encapsulation to check the functionality. Our biosensors showed a detectable response both in media/buffer and in hydrogel materials. The results demonstrate that our engineered biosensors are suitable to be turned into adhesive patches to produce outputs which can be visualized and analysed via image processing applications corresponding to biomarker concentrations in sweat.

This work has been supported by ONR Global Grant #N62909-20-1-2036.

Ultrasonic Reporter for Kinase Activity (UReKA)

Jee Won Yang

Although protein kinases control nearly every facet of cellular function, current approaches to understanding their tightly controlled spatiotemporal regulation of multiple cellular processes are limited to *in vitro* assays due to the lack of robust, deep-tissue *in vivo* biomolecular tools. As aberrant kinase signaling is linked to numerous pathologies, novel tools for imaging live kinase activity are increasingly needed to study the rapid dynamics of and complex mechanisms behind aberrant signal transduction pathways. While advances in fluorescent proteins and live-cell optical imaging have allowed spatiotemporally precise kinase activity measurements in living cells, they do not enable noninvasive deep-tissue imaging due to the physical limitation of light scattering.

Alternatively, ultrasound has unique advantages for cellular response imaging due to its ability to penetrate much deeper into tissue (several cm) than light with relatively high spatial and temporal resolution (<100 µm and 1 ms). Additionally, genetically encoded contrast agents for ultrasound-based on gas vesicles (GVs), a unique class of air-filled protein nanostructures derived from buoyant microbes, have been developed as acoustic reporter genes and as acoustic biosensors for enzymatic activity. Here, we have engineered GVs as the first ultrasonic reporter for kinase activity (UReKA), enabling dynamic monitoring of Protein Kinase A (PKA) activity reversibly and repeatably when UReKA is administered. The hypothesis is that upon phosphorylation by PKA, the outer structural proteins of UReKA undergo an allosteric conformational change with a loss of binding to GV shells, resulting in reduced GV shell stiffness, increased mechanical deformation behavior (e.g., buckling), and enhanced detection of a non-linear signal. Future experiments include demonstrating the functionality of UReKA *in vitro* in mammalian cells and *in vivo* to measure upregulated PKA activity.

Short Activators and Repressors of RNA Toehold Switches

Megan McSweeney, Alexandra Patterson, Monica McNerney, Kathryn E. Loeffler, Nikki McArthur, Ravindra Kane, Mark Styczynski

Cell-free expression (CFE) systems are a promising platform for engineering diagnostic biosensors that are affordable, simple to use, and interpretable without any equipment. CFE biosensors have been engineered to detect a variety of clinically relevant nucleic acids, ions, and small molecules using generalizable sensing schemes. Nucleic acid biosensors commonly use *de novo* designed riboregulators, while ion and small molecule sensors typically use transcription factor-based approaches. Despite their generalizability, these existing technologies neglect an important class of biomarkers: proteins. An efficient way to detect proteins using CFE would fill a crucial gap in the field and enable rapid diagnosis of diseases that currently require expensive antibody-based testing. Here, we discuss the first modular CFE biosensing platform for generalizable protein detection that we call TLISA (**T**7 RNA polymerase-**L**inked **I**mmunosensing **A**ssay). TLISA uses a proximity-dependent split T7 RNA polymerase with protein-binding domains fused to each fragment that bind to the protein of interest. When the protein of interest is present, both polymerase fragments bind to it, reassemble, and turn on reporter gene expression. We show that TLISA is robust to the use of many different nanobody and monobody fusions, making this platform easily adaptable to sense different target proteins. TLISA functions in human serum and can yield results in as little as 25 minutes, which demonstrates practicality for use in point-of-care (POC) settings. TLISA also responds in a dose-dependent manner, allowing for semi-quantitative result readout – a critical feature that state-of-the-art POC lateral flow assays typically lack. We also showed that TLISA can be used to engineer clinically relevant biosensors by constructing sensors for transthyretin (a serum biomarker for malnutrition) and for SARS-CoV-2. Altogether, TLISA represents an impactful new tool for synthetic biologists that offers modular cell-free protein sensing for the first time, enabling rapid detection of virtually any protein biomarker.

Biofoundry-Assisted Expression and Characterization of Plant Proteins

Quentin Dudley

Many goals in synthetic biology, including the elucidation and refactoring of biosynthetic pathways and the engineering of regulatory circuits and networks, require knowledge of protein function. In plants, the prevalence of large gene families means it can be particularly challenging to link specific functions to individual proteins. For example, enzymes in tobacco are responsible for unwanted glycosylation/oxidation of a heterologous pathway to produce the monoterpene indole alkaloid cancer drug vinblastine. However, protein characterization has remained a technical bottleneck, often requiring significant effort to optimize expression and purification protocols. To leverage the ability of biofoundries to accelerate design–built–test–learn cycles, we present a workflow for automated DNA assembly and cell-free expression of plant proteins that accelerates optimization and enables rapid screening of enzyme activity. We built a DNA assembly toolbox, a luciferase-based system for quantification, and demonstrate functional assays which can be performed without the need for protein purification. Together, the combination of automated assembly of DNA parts and cell-free expression reactions should significantly increase the throughput of experiments to test and understand plant protein function and enable improved downstream plant engineering workflows.

Toehold Switch Plus Signal Amplification Enables Rapid Detection

Kevin Morey, Al Watson, Jacob Sebesta, Christie Peebles

Recent world events have led to an increased interest in developing rapid and inexpensive clinical diagnostic platforms for viral detection. Here, the development of a cell-free toehold switch-based biosensor, which does not require upstream amplification of target RNA, is described for the detection of RNA viruses. Toehold switches were designed to avoid interfering secondary structure in the viral RNA binding region, mutational hotspots, and cross-reacting sequences of other coronaviruses. Using these design criteria, toehold switches were targeted to a low mutation region of the SARS-CoV-2 genome nonstructural protein 2. The designs were tested in a cell-free system using trigger RNA based on the viral genome and a highly sensitive fluorescent reporter gene, mNeonGreen. The detection sensitivity of our best toehold design, CSU 08, was in the low picomolar range of target (trigger) RNA. To increase the sensitivity of our cell-free biosensor to a clinically relevant level, we developed a modular downstream amplification system that utilizes toehold switch activation of tobacco etch virus (TEV) protease expression. The TEV protease cleaves a quenched fluorescent reporter, both increasing the signal fold change between control and sample and increasing the sensitivity to a clinically relevant low femtomolar range for target RNA detection.

RNA-Toehold Switch Technology Enables Customizable Biological Threat Detection

Kurt Gust, Xiao Luo, Michael Rowland, Ping Gong

Ensuring warfighter safety relies on detecting biological threats in the battlefield environment. RNA-toehold switch (RTS) biotechnology offers a customizable tool set for detecting active biological agents through species- and/or strain-specific RNA expression. By binding to the target sequence in the RTS, the RNA of a threat agent induces a conformational change, allowing expression of a custom reporter like green fluorescent protein (GFP). To develop effective RTS for threats like COVID-19 and *Staphylococcus aureus*, we have designed novel RNA targets and assessed switch performance. DNA templates of RNA targets were initially synthesized, cloned into bacterial plasmids, and transformed into competent cells to generate high copy numbers of the plasmids for testing. Quality assurance was performed on the plasmids by PCR and DNA sequencing, after which the RTS was amplified using T7 promoter primers to ensure sufficient copies for cell-free assays after purification. Synthetic DNA and RNA samples were used to characterize the RTS and optimize detection. Currently, the COVID-19 RTS is being tested on paper discs for potential transition to a handheld threat detection system developed by the US Army DEVCOM. The *S. aureus* RTS has been validated for genus-level detection, and a complementary, strain-specific RTS is being developed to selectively monitor the expression of the MecA gene in the methicillin-resistant variants of *S. aureus*. This effort has provided proofs of concept for the RTS biotechnology to identify critical biological threat signatures, and ongoing work aims to transition and test this synthetic biotechnology for field-based biological threat detection using cell free systems.

Paper-Based, Cell-Free Detection of Heavy Metals for in-Field Water Quality Testing

Casey Bernhards, Kathryn Beabout, Meghna Thakur, Eric Brack, Michael Wiederoder, Scott Walper, Gregory Ellis, Jorge Chávez, Matthew Lux

The ability to detect heavy metals and other contaminants in water is critical for health and safety in both field-forward and resource-poor environments. The power of biology is being leveraged to develop biosensors capable of responding to a wide variety of analytes; however, deployment of these sensors remains a challenge. Biosensors based on cell-free expression (CFE) systems hold great potential for providing rapid, in-field detection capabilities. CFE systems can be freeze-dried onto paper substrates, and following rehydration at the point of need, a visualizable reporter protein is produced in response to the target analyte. These paper-based sensors are cheap, disposable, simple to use, and multiplexible. This work shows a tri-service effort to develop cell-free biosensors for arsenic, cadmium, and mercury. Under laboratory conditions, all three sensors are able to detect heavy metals at or below Military Exposure Guideline (MEG) levels, and the arsenic and mercury sensors achieve detection limits below the World Health Organization (WHO) recommended levels within 30 minutes. To further test and optimize the sensors for field-relevant conditions, the sensors were evaluated using an NSF test water standard and after lyophilization on paper tickets. Different optimization strategies were employed to improve the performance of each sensor under these conditions, and detection of the three heavy metals at MEG levels was demonstrated in spiked environmental water. In addition to heavy metals, the CFE approach described here can be expanded in the future to detect a wider range of contaminants, including toxic compounds and pathogens. Ultimately, a panel of different CFE sensors could be combined onto a single paper ticket, allowing for the simultaneous detection of multiple hazardous water contaminates in the field.

A Strategy to Improve Expression of a Type III Polyketide By Refactoring Genetic Parts Via Lysate-Based Cell-Free Expression

Tien Miah Sword

Recombinant proteins are important tools in microbial systems as they have a wide range of applications in the biomanufacturing industries. However, optimal conditions to express a protein of interest heterologously do not exist. Therefore, it is critical to establish a production system for proteins in a manner that is high-yield and cost-effective. Choices and design of plasmids play an important role in minimizing metabolic burden and increasing yield for recombinant proteins. Hence, to achieve a more effective system, an important parameter to consider is characteristic of genetic elements (e.g. coding sequence and/or promoter). Besides time-consuming procedures, the bottleneck of conventional cell-based methods of protein expression often is cell toxicity. To bypass this, cell-free expression (CFE) platforms derived from crude cell extracts can be used to prototype plasmid constructs for *in vivo* enzyme production, expediting genetic parts characterization from weeks to days. CFE is a simple open system that allows direct manipulation of transcription, and translation, without the use of living cells; hence, it is easier to control via varying parameters. While proven for robustly synthesized proteins and T7 promoters, lysate-based CFE's utility for prototyping expression of more complex proteins, such as biosynthetic pathways remains unclear. Here, the performances of constructs expressing a type III PKS, RppA, with varying DNA regulatory elements are evaluated in *E. coli* lysates and cells. RppA produces a red pigment that can be used as a reporter for high-throughput characterizations. First, we focus on testing parts that may benefit PKS expression, then we test correlative performances between *in vivo* and *in vitro* systems. Modular BioBrick vectors carrying non-T7 inducible promoters, an uninducible promoter, and different RppA coding sequences will be used. Establishing the utility of lysates for prototyping PKS expression will accelerate the construction of valuable polyketide-forming microbial cell factories, diversifying access to natural product scaffolds.

Engineering Receptors to Control Platelet Activation and Therapeutic Release

Shwan Javdan, Tara L. Deans

While the use of living cells as therapeutic delivery vehicles has become a prominent area of focus in recent years, challenges remain in achieving clinical translation due to limitations on available therapeutic loading methods as well as lack of controlled therapeutic release mechanisms. Platelets offer a unique delivery platform as they are anucleate cells naturally loaded with proteins from their megakaryocyte precursor cells, and release their intracellular contents once activated. We have shown that megakaryocytes can be loaded with proteins that are packaged into platelets, creating engineered platelets that function as novel delivery devices. However, a key challenge remains that the native activation pathway of platelets may cause the activation and subsequent release of therapeutic payloads in undesired locations. Thus, there is a critical need for a controllable activation trigger in engineered platelets, specifically, one that is orthogonal to the native activation pathway. Here, we engineered a set of novel chimeric platelet receptors that can be activated only by a pharmacologically inert small molecule drug to enable the controlled release of the therapeutic payload carried by engineered platelets.

CRISPR-Based Engineering Tools to Control Individual Microbes and Microbiota without Antibiotic Resistance Genes at a Single Strain Level

Austin Rottinghaus, Matthew Amrofell, Steven Vo, Tae Seok Moon

Microbial biocontainment is essential for engineering safe living therapeutics [1, 2]. However, the genetic stability of biocontainment circuits is a challenge. Kill switches are among the most difficult circuits to maintain due to the evolution of escape mutants. We engineered two CRISPR-based, chemical- or temperature-inducible kill switches in the probiotic *Escherichia coli* Nissle and demonstrated mutationally robust biocontainment [3]. In this presentation, we will discuss our machine learning-based microbiota engineering tools that are useful to manipulate microbiota and kill pathogens at a single strain level [4]. Specifically, we will discuss the development and validation of a novel computational program, ssCRISPR, which designs strain-specific CRISPR guide RNAs (gRNAs) that can be utilized to modify complex consortia. As a proof of concept, we applied the program to two novel applications: the isolation of specific microbes from consortia through plasmid transformations and the removal of specific microbes from consortia through liposome-packaged CRISPR antimicrobials. Additionally, we will discuss antibiotic resistance gene-free plasmid systems that prevent antibiotic resistance spread via horizontal gene transfer [5]. This new technology has vast implications in designing strain-specific antimicrobials and combating the growing concern of antibiotic- and bacteriocide-resistant microbes.

1. Curr Opin Biotechnol, 2020. 66, 11-17.
2. Biotechnology Journal, 2020. 15, 1900319.
3. Nature Communications, 2022. 13, 672
4. PNAS. 2023. 120, e2213154120
5. Trends. Biotechnol, 2022, 40, 1405-1414

Engineering the Field Microbiome for Extreme Remote Biosensing Using Hyperspectral Reporters

Yonatan Chemla, Christopher A. Voigt

Living biosensors hold great promise for detecting molecules crucial for agriculture, health, and defense, including soil/water pollutants, human disease markers, and landmine residue. Bacterial biosensors can be affordably scaled up and easily deployed to cover large areas that need to be monitored with high resolution over time, making them ideal for field applications. However, current reporter technology has a poor ability to extract information from deployed biosensors and thus prevents these applications from reaching maturity. Despite their widespread usage and popularity, both fluorescence and luminescence spectroscopies and traditional colorimetric analysis possess limited utility for remote sensing outside of a laboratory setting due to specialized equipment, background interference, expensive substrates, inability to accumulate over time, inability to diffuse out of the cell and the burden they impose on the host. In this talk/poster, I will describe our unpublished results in developing a new field reporter modality based on unique light-absorbing small molecules and hyperspectral imaging. We developed a quantum-based computational pipeline to predict the spectral signatures of all known metabolites and identified molecules with unique spectral signatures as candidate reporters. The network of all reported enzymatic reactions was analyzed to determine the most readily synthesizable candidate reporters in various bacteria. We have engineered and tested these genetic pathways in *E. coli*, *B. subtilis*, and a non-model soil bacterium – *R. gelatinosus*. We have shown that the chassis can sense molecules under lab and field conditions and transduce the information to extreme distances using drone and satellite-mounted hyperspectral cameras and a simple image analysis algorithm.

Single-Cell Phenotyping of Extracellular Electron Transfer (EET) Via Microdroplet Sorting of Cu(I)-Catalyzed Alkyne-Azide Cycloaddition (CuAAC)

Gina Partipilo, Emily Bowman, Hal Alper, Benjamin Keitz

Bacterial consortia found in the human gut, soil, and natural bodies of water can often prove challenging to study in a laboratory setting. Removing samples from their native environment and placing them into growth media can cause resource competition, fewer native niches, and a decrease in microbial diversity. Herein, we describe the use of a generalizable droplet microfluidic system and Cu(I)-catalyzed alkyne-azide cycloaddition (CuAAC) to screen for bacteria with the ability to perform Extracellular Electron Transfer (EET). While microfluidic systems have been used in the past to study bacterial consortia, many focus on sorting out metabolite production. Our work seeks to screen for specific phenotypes. First, we establish an oxygen-limited pico-injection method for performing CuAAC inside the microfluidic system. Then, we use the method to enrich wild-type *S. oneidensis* MR-1, a known EET-capable bacteria, from a *S. oneidensis* knockout strain lacking key EET-proteins located in the membrane. Then, the screen was implemented to enrich *S. oneidensis* MR-1 from a synthetic mixed population comprised of *E. coli* Nissle 1917, and *S. cerevisiae* BY4741. Finally, bacteria harvested from mineral sediment from Lake Austin, was submitted though the screen in an attempt to identify novel EET activity compared to a traditional batch screening protocols which have had success previously. Overall, this work represents a new oxygen-limited method of screening for a specific phenotypic activity in a mixed microbial population.

A Novel Secretory Protein Production System for *E. coli* Nissle to Deliver Therapeutical Proteins

Recep Erdem Ahan, Cemile E. Ozcelik, Irem Niran Cagil, Urartu Seker

E. coli Nissle 1917 is a clinically proven bacterial chassis engineered for the treating different metabolic diseases. Although, targeted protein delivery seems plausible with EcN with repurposing natural secretion systems, inherent drawbacks of natural secretion systems such as high metabolic load, narrow substrate profile, low versatility and their attribution to pathogenicity limit the development of EcN as a protein delivery shuttle. An efficient protein secretion can enhance capabilities of EcN for this purpose. Here, we present a new-to-nature secretion system for EcN called iLOM-SS, acronym for **i**nducible **L**eaky **O**uter **M**embrane based **S**ecretion **S**ystem. In iLOM-SS, proteins are translocated from cytosol to periplasm via sec targeting signal peptide and released to the extracellular space by translational repression of a protein whose knock-out mutant leads to leaky outer membrane phenotype. We constructed iLOM-SS in EcN and showed that proteins up to 90kDa can be secreted with the proposed system. Further characterizations revealed that iLOM-SS is modular and fast protein secretion system which offers titratable protein release. As a proof-of-concept, a human anti-inflammatory cytokine, IL-1Ra, was secreted via iLOM-SS in its active form. In addition, the leaked periplasmic content from EcN was proven to cause no cytotoxic and proinflammatory effects *in vitro* on Caco-2 cells. We foresee that iLOM-SS will enable to engineer the EcN strain for therapeutical interventions wherein protein delivery is necessary.

Trojan Horse RNA Antimicrobials That Are Selectively Activated Using Native mRNA

Maria Claudia Villegas Kcam, Kim Wai Mo, Jonathan Silberg, James Chappell

The rise in antimicrobial-resistant pathogens is a major threat to public health and is predicted to result in >10 million deaths annually by 2050. Conventional chemical antibiotic discovery pipelines are depleted, and current antibiotics kill both commensals and pathogens, thereby massively disrupting our microbiomes, which support our health. Further, this non-selective killing has led to an explosion of antimicrobial resistance, which represents >$1B in health care costs annually. To overcome these challenges, we developed a new class of genetically-encoded RNA therapeutics that serves as a highly designable and selective antimicrobial. Our platform, called RNA-Activated Protein by Trans-splicing Ribozyme (RAPTR), is a programmable RNA-based system that encodes for a silent 'death gene' that is selectively activated by pathogen-specific target mRNA. As the RNA input and protein outputs are modular, this platform is facile to redesign to new threats. Here, we will present our work on creating prototype RAPTRs that kill bacterial threats with high efficiencies (99.99% killed) using multiple action mechanisms (DNA cleavage and inhibition of DNA replication). To understand RAPTR, we applied a high-throughput selection to large libraries of RAPTR variants that target diverse native and non-native mRNAs. We show that RAPTR can function with diverse transcripts, which vary in their cellular copy numbers and have translation initiation sites of varying strengths. Ongoing efforts are investigating how RAPTR functions within consortia containing and lacking clinically-derived pathogens. In summary, RAPTR provides a reprogrammable, versatile, and selective RNA therapeutic that can address current unmet needs for infectious diseases.

A High-Throughput Analysis of Genetic Influence in Microbial Growth

Zachary Holmes, Lingchong You

Microbial growth patterns are important to understand because they contain information relating to characterization of different phenotypes, understanding how microbes grow under different conditions, and can help select the correct microbes for given applications.

Properly characterizing microbial growth patterns is not a trivial task. Some limitations include equipment being unable to detect the microbes at meaningful time points, knowing the exact starting number of cells, and error propagation through calculations.

We present a high-throughput method to characterize microbial growth patterns, which is independent of equipment used and robust to error propagation. We used robotic laboratory equipment to generate 360 growth curves in a single 24-hour experiment. We standardize the measurement of microbial growth rates by inoculating cultures at two different quantities and looking at the time delay between the two growth curves (Youmans and Youmans, 1949). With an accurate maximum growth rate, we can better characterize the lag time of the microbes.

Using our method, we have characterized strains with different genetic backgrounds to determine how different genetic backgrounds influence growth patterns. These results have shown that some categories of gene knockouts result in changing growth rates, lag times, and final microbe densities.

We apply deep-learning models to reduce the dimensionality of the data. This allows us to cluster and characterize microbes with different genetic backgrounds, and we can use this information to predict growth patterns of microbes.

Directing Lachnospiraceae Biosynthesis with Novel Genetic Tools

Jack Arnold

Recent correlative studies have linked the metabolism of the microbiota to the onset or prevention of various disease states. Complementary advancements in 'Omics approaches and anaerobic culture techniques have significantly de-risked harnessing diverse native microbiota members as chassis for implementing synthetic biology strategies. Of great import, gut symbionts, namely butyrate-producing Lachnospiraceae, are thought to play disease-protective roles and provide a unique access point for interventions that modulate host physiology without directly targeting the mammalian host. However, our explicit understanding of how these microbes affect host-facing protective mechanisms remains elusive. This knowledge gap is due to a continued dearth of genome manipulation technologies to control the metabolic activity of health-associated gut bacteria. Here, we propose a blueprint to enable genetic engineering in currently intractable symbionts of the Lachnospiraceae family. Establishing these genetic tools would allow for direct genotype-to-phenotype analyses of important "beneficial" gut symbionts. Additionally, with new insight into the physiology of these microbes, we can both genetically augment living therapeutics and uncover novel microbial metabolites to treat gut-associated disease. To this end, we co-opted plasmid-based systems to establish basic genetic engineering capabilities in Lachnospiraceae. These tools served as a prototyping platform to develop a robust genetic toolkit that enables precise manipulation of gene expression in Lachnospiraceae. Moving forward, we will employ these strategies to regulate the production of potentially beneficial native or synthetic metabolites. The development of these tools will aid precise genetic study of many key symbiotic microorganisms, provide a basis to genetically manipulate other intractable microbes, and allow for tuning microbial function in defined contexts to investigate hypotheses that have remained beyond our reach.

DNA Editing Game (DEGA) Theory

Nicholas Roehner

DNA editing technologies such as CRISPR/Cas-based systems hold great promise for treating diseases, enhancing agricultural crops, and enabling biological information storage, but this promise is tempered by the potential for their unchecked misuse. Thus far, biodefense and biosecurity concerns for these technologies have largely focused on containment of their resulting modifications or safeguarding their implementation as autonomous gene drives. As DNA editing becomes more effective and more widely deployed, however, it will become necessary to consider the security of DNA sequences being edited by multiple actors with conflicting goals. While there currently exist computational methods and tools to assess potential sites for DNA editing using different systems, few if any approaches have been developed to evaluate these systems in a head-to-head contest. To address this gap in biodefense and biosecurity capabilities, we have applied combinatorial game theory to estimate the edit-based advantage that a DNA editing system or set of systems has relative to an opponent's. Unlike existing tools for site identification, our approach takes into account that some edits can prevent others from occurring, for instance by disrupting the PAM site required by a CRISPR-Cas based system. Investigating one-versus-one matchups between CRISPR base editors, we have identified 5 out of 18 with predominately winning records in rice and bread wheat. Differences in PAM site variability appear to have the largest effect on a winner's edit advantage, compared to other factors such as edit window length and choice of deaminase. Going forward, we can use our approach to inform the design of DNA editing systems with superior edit advantage, as well as the design of DNA sequences that are less vulnerable to DNA editing.

Engineering Bacterial Swarm Patterns for Robust Encoding and Decoding of Environmental Inputs

Anjali Doshi, Marian Shaw, Ruxandra Tonea, Soonhee Moon, Rosalía Minyety, Anish Doshi, Andrew Laine, Jia Guo, Tal Danino

A diverse array of bacteria species naturally self-organize into durable macroscale patterns on solid surfaces via swarming motility—a highly coordinated, rapid movement of bacteria powered by flagella. Engineering swarming is a promising opportunity to increase the scale and robustness of coordinated synthetic microbial systems. Here we engineer *Proteus mirabilis*, which natively forms centimeter-scale bullseye swarm patterns, to "write" external inputs into visible spatial records. We previously engineered tunable expression of swarming-related genes that modify pattern features in response to single inputs and two inputs simultaneously, and developed quantitative approaches to decoding with deep learning. We transition from classification-based decoding to regression-based decoding of concentrations from our dual-input strain, which can increase future applicability by requiring less stringent conditions when acquiring training data. We next characterize the robustness of our strains' IPTG-readout at changing temperatures. With an assay protocol modified for greater ease-of-use, we also validate the robustness of the readout in the presence of minimally-filtered natural water samples. We then engineer strains of *P. mirabilis* which sense the presence of copper, zinc, and mercury, and characterize their response via GFP-expression. Finally, we engineer a strain of *P. mirabilis* which records the presence of aqueous copper with visible changes to its ring pattern. Altogether, this work furthers the proof-of-concept for the application of engineered swarm patterns as environmental records, expanding the framework for engineering emergent microbial behaviors for biotechnology.

Design and Computational Analysis of Synthetic Circuits with a Non-Standard Dead-Cas9 System in Engineered Bacteria

Lorenzo Pasotti, Davide De Marchi, Roman Shaposhnikov, Maria Gabriella Cusella De Angelis, Paolo Magni

Increasing the success in the design of synthetic circuits is highly important to limit the number of trial-and-error steps in biological engineering. Key efforts in this direction include the expansion of the toolkit of orthogonal components, and the improvement of descriptive and predictive mathematical models. In this work, we showcase the potential in bacterial synthetic biology of the Staphylococcus aureus dead-Cas9 (SadCas9), a low-size high-specificity CRISPRi system not previously adopted to construct information processing circuits in bacteria [1,2]. Logic gates, feedforward loops and transcriptional cascades expressing reporter proteins as outputs showed a qualitatively expected behavior. However, quantitative analysis of steady-state data elucidated unexpected features in curve shapes, requiring the support of computational tools to debug the circuits. Empirical and biomolecular kinetic models were defined that theoretically captured the unexpected trends by considering cell load and small copy numbers of the regulatory proteins. We anticipate that the detected trends can affect different circuit architectures and other components, possibly leading to wrong mechanistic conclusions during circuit analyses or to the malfunctioning of bottom-up engineered circuits. The study of similar circuit collections will greatly support the development of accurate modeling tools for biological engineering, and the use of programmable CRISPRi components like SadCas9 will enable to scale-up the design of complex circuits.

[1] Ran FA et al. In vivo genome editing using Staphylococcus aureus Cas9. *Nature* 2015, 520:186-91.

[2] Friedland AE et al. Characterization of Staphylococcus aureus Cas9: a smaller Cas9 for all-in-one adeno-associated virus delivery and paired nickase applications. *Genome Biol* 2015, 16:257.

Metataxonomic Memory: Consortia-Scale Information Storage within the 16S rRNA of an Environmental Microbiome Using a Ribozyme

Prashant Kalvapalle, Lauren Gambill, August Staubus, Matthew Dysart, Kiara Reyes Grandel, Li Chieh Lu, James Chappell, Lauren Stadler, Jonathan Silberg

Biological memory systems have enabled the storage of information reflecting changes in physical and chemical conditions, or signatures of incoming foreign DNA sequences in the DNA of an organism. Until now, biological memory systems have largely been limited to model organisms such as *Escherichia coli*. We have extended the information storage paradigm into complex natural microbiomes where information is stored in the 16S rRNA of the consortia members. We achieve this information storage using a ribozyme that targets a conserved region of 16S rRNA, and splices a barcode in that region. The readout of the barcode along with the neighboring host 16S variable regions are performed by RT-PCR followed by metagenomic amplicon sequencing. As a proof of concept, we apply this RNA-addressable memory (RAM) to map the host range of a plasmid undergoing horizontal gene transfer. This plasmid has RAM repressed in the donor and is de-repressed upon conjugation, thereby activating the splicing reaction only in the successful recipients. We demonstrated the RAM system by using an E. coli donor to conjugate the plasmid into a wastewater microbiome. We detected the barcoding of 43% of all observed taxa in the population (120 out of 280 ASVs), spanning 6 taxonomic orders, confirming that RAM works in a diversity of transconjugant organisms. By normalizing the read counts for the barcoded vs un-modified 16S rRNA, we identify ASVs with conjugation 'efficiency' varying across 3 orders of magnitude. We also detect barcoding from rare organisms with undetectable 16S within the original community demonstrating the sensitivity of the method to detect rare transconjugants inaccessible by any other methods. As a new paradigm of RNA based information storage, ribozymes coupled with conditional promoters can enable detection and storage of many other forms of information in the microbiome to take synthetic biology into complex environmental microbiomes.

Patterned Engineered Living Materials Based on *Escherichia coli* Biofilms

Roberto Avendaño, Javier Santos-Moreno, Yolanda Schaerli

Bacterial biofilms possess properties that are highly desirable for the next generation of materials: they emerge through self-organization controlled by a genetic program, they can self-repair and adapt to environmental cues, and they are driven by sustainable energy sources and biodegradable. However, in order to realize the full potential of these materials, we need to be able to control their spatiotemporal organization. The aim of this project is to establish synthetic circuits to control the spatial expression of curli fibers in of *Escherichia coli* biofilms to produce patterned functionalized living materials. To this end, we use a synthetic toggle switch, composed of two mutually inhibitory nodes. Controlled by diffusible molecules it can produce discrete stripes of gene expression. So far, we successfully coupled the curli production with the toggle switch patterning, so that each node creates a specific type of curli fiber and we are currently functionalizing our patterned engineered living biofilm material.

Manipulating DNA Fate in Plant Cells after *Agrobacterium*-Mediated Delivery with an Integrase-Based Tool

Jennifer A.N. Brophy, Vivian Zhong

Transkingdom delivery of DNA from *Agrobacterium* to plants is a stunning biological feat and the basis of powerful plant biotechnology tools. Although it is the predominant mechanism by which new DNA is introduced to plants by engineers, *Agrobacterium*-mediated plant transformation integrates transfer DNA (T-DNA) into the genome at random loci. This random integration can interfere with the expression of native genes and leads to highly variable transgene expression. The ability to control T-DNA integration would enable more precise and quantitative plant engineering. Unfortunately, the mechanism of T-DNA integration remains poorly understood and is therefore difficult to control.

We developed a plant line in which T-DNA integration can be easily studied. This plant contains an integrase-invertible promoter positioned upstream of a reporter gene such that T-DNAs encoding the phiC31 integrase activate expression of the reporter when they are successfully delivered to the plant cell. Using this system, we found that integration occurs in less than 50% of successful delivery events. We seek to use this tool to measure the frequency and timing of T-DNA integration during embryo development and to test modifications that tune integration frequency. By combining the integrase-based reporter of T-DNA delivery with CRISPR/Cas9, we are developing a system that facilitates the rapid generation of Cas9-free edited plants. This would streamline plant engineering by enabling genome editing without the possibility of disrupting off-target genetic elements through T-DNA integration.

Machine Learning Guided Engineering of CYP76AD1 Using Ultra High-Throughput Microfluidic Enzyme Screening

Mark Mahnke

Engineered enzymes occupy a critical role in modern biotechnology and medicine. Despite their importance, the engineering process remains a time-consuming and resource-draining process due to unknown and rugged nature of the sequence-function landscape. High-throughput screening used *in tandem* with machine learning helps alleviate these issues through the rapid acquisition of data and extraction of important mutations from that data. Here these principles are applied to the P450 enzyme CYP76AD1 that catalyzes the conversion of L-tyrosine to L-DOPA, an important clinical small molecule. By employing the enzyme DOPA dioxygenase (DOD) L-DOPA is converted to betaxanthin, a fluorescent and highly water-soluble small product that is used for the indirect detection of L-DOPA production. Here, we generated a mutant library of CYP76AD1 variants in *E. coli* and encapsulated them in a microfluidic system where we screened millions of variants at a rate of 2000 drops per second, generating multiple data points for each variant's activity. We use positive-unlabeled learning (PU-learning), a unique machine learning model that capitalizes on unlabeled data, to learn the sequence-function relationship from next generation sequencing data and identify important mutations from our screen. 5 mutants were selected and tested against the wildtype and we attempt to elucidate the biophysical mechanisms that underly their performance.

Placeholder Title.

Cyrus Knudsen, Markus Covert

The first Boeing 777 jet was first built entirely from in silico designs, and was subsequently tested by just fueling it up and flying it. A cell simulation model that could be used as a "digital twin" to prototype new pathways and genetic circuits would rapidly improve the rate at which new solutions can be engineered. The E. coli Whole-Cell Modeling Project aims to develop a simulation model for this purpose.

One challenging aspect of constructing whole-cell simulation models is the estimation of a consistent set of parameters given a heterogeneous set of data. With measurements of kinetic parameters, substrate concentrations, equilibrium constants, all with various degrees of uncertainty, how can we make the most of our data while still ensuring stable cell growth? If we introduce a new metabolic pathway, can we quickly determine its dynamics with a few measurements?

In this work, we demonstrate a general algorithm based on convex optimization to make quantitative estimates of kinetic parameters, regulatory parameters and metabolite concentrations for a biochemical reaction/regulation network that fit a given set of reaction fluxes. The method is robust to any amount of missing data, allows for quantifying parameter uncertainty and includes metabolite-determined thermodynamic constraints.

Using state-of-the-art convex optimization libraries, runtime is less than 0.1 seconds for networks with tens of reactions, and in the range of seconds for networks with hundreds of reactions. We demonstrate the performance on both a small example problems and the entire metabolic network of *E. coli*, provided with enzyme and flux values from the current *E. coli* Whole-Cell Model.

Engineering Acoustic Reporter Genes for Multiplexed Ultrasound Imaging

Nivin Nyström, Zhiyang Jin, Mengtong Duan, Przemysław Dutka, Ruby Zhang, Dina Malounda, Mikhail Shapiro

Introduction. Acoustic reporter genes (ARGs) that encode gas-filled protein nanostructures called gas vesicles (GVs) have permitted deep-tissue imaging of engineered cells at high spatial resolutions via ultrasound. Whereas initial imaging methods required high pressures to collapse GVs, followed by image subtraction to identify their locations, real-time imaging with ARGs was recently accomplished by applying sub-collapse pressures that caused GVs to undergo reversible acoustic buckling, which specifically generates nonlinear acoustic signals on ultrasound. Engineering ARGs with distinct nonlinear acoustic spectra that can be deconvoluted from one another would enable multi-variable tracking of cells and/or molecular events, analogous to color-shifted proteins for fluorescence imaging. **Objective.** We therefore wanted to develop ARGs that generate distinct nonlinear acoustic spectra for multiplexed imaging of cells *in vivo* on ultrasound. **Methods and Results.** GVs consist of a major structural protein called gvpA that makes up their shell, and a minor structural protein gvpC that scaffolds the shell for increased rigidity. We designed a high-throughput assay based on the light-scattering property of GVs to identify conditions for GV assembly in *E. coli* independent of gvpC co-expression. The ΔgvpC construct exhibited a 3.1-fold increase (n=3, $p<0.0001$) in maximum nonlinear acoustic signals relative to the wild-type construct, allowing for more sensitive real-time imaging of bacteria on ultrasound. Next, we generated a mutant library for gvpC using error-prone PCR and screened through the nonlinear acoustic spectra of our mutants using a custom-built motorized ultrasound system. Of the mutants that generated nonlinear acoustic signals (N=322), we identified and validated one mutant ($gvpC_{mut}$) that exhibited a pressure-shifted nonlinear acoustic spectrum (ΔP = 540 kPa) relative to ΔgvpC, with near-equivalent maximum nonlinear acoustic signals relative to ΔgvpC. **Conclusion.** Our immediate next steps include multiplexed imaging of tumor-homing bacteria in mice using the ΔgvpC and $gvpC_{mut}$ constructs, and translation of $gvpC_{mut}$ into mammalian cells.

Engineering Extracellular Electron Transfer in *Escherichia coli* for Microbial Electrochemical Devices

Mohammed Mouhib, Melania Reggente, Lin Li, Nils Schuergers, Ardemis Boghossian

Microbes hold great potential for green electrochemical synthesis as well as bio-electricity generation from organic waste. The key to the successful development of microbial electrochemical devices lies with the ability of the microbes to transport electrons across their outer membranes. These devices thus often rely on the use of microbes known as exoelectrogens, which include bacteria that have evolved metabolic pathways that enable extracellular electron transfer (EET) to solid substrates for cellular respiration. While exoelectrogens demonstrate facilitated electron transfer with the electrodes of these devices, natural exoelectrogens are often restricted to utilizing a narrow range of substrates that limits their versatility for processing different organic wastes and electrosynthesizing a rich diversity of high-value chemicals. By contrast, non-exoelectrogenic microbes such as *Escherichia coli* that lack efficient EET pathways are host to rich metabolic reaction networks that can be more readily tailored for various applications using synthetic biology.

In this work, we biologically engineer *E. coli* for enhanced EET. To this end, we expressed different combinations of proteins from the natural exoelectrogen *S. oneidensis* MR-1, as well as *E. coli* native proteins to optimize EET of the bioengineered strain. We further introduced biosynthesis pathways for redox mediators to facilitate EET. The bioengineered strains show significant improvements in electron transfer rates, as confirmed through colorimetric reduction assays of soluble electron acceptors and electrochemical characterizations following electrode reduction. The improved EET demonstrated in this work paves the way for increasing the efficiency of existing *E. coli*-based electrochemical devices while opening the doors to new applications that benefit from the broad chemical repertoire of these microbes.

Creating Synthetic Origins of Replication (SynOri) through Refactoring and Re-Engineering of the pMB1 Origin

Baiyang Liu, James Chappell

Plasmids remain by far the most common medium for delivering engineered DNA to organisms. However, the current menu of plasmid systems is limited and imposes significant practical drawbacks that include fixed copy numbers that only sparsely cover the desirable range, a limited number of compatible plasmids, and a reliance on natural replication mechanisms that are inherently non-modular, which complicates plasmid engineering. In this study, we aim to construct synthetic origins of replication (SynOri) through refactoring and reengineering the naturally occurring pMB1 origin of replication. First, we open the "black box" of this natural replication mechanism by genetic refactoring to gain insights into its genetic architecture and the nature and operation of the negative feedback loop that achieves copy control. By manipulating the relative expression strength of these components, we can create a library of plasmids with customizable copy numbers. We then create new modular synthetic origins that use synthetic RNA transcriptional repressors to implement copy control. Importantly, this allows for the creation of orthogonal SynOris whose copy number can be independently controlled. We further demonstrate that the implementation of synthetic origins of replication introduces a novel layer of gene regulation based on plasmid copy number, that could allow for the development of inter-plasmid circuits and plasmid-copy-based biosensors. Overall, this study establishes the feasibility of creating plasmids with synthetic origins of replication that can serve as a new biotechnology for synthetic biology.

Continuous Recording of Cell History into DNA

Theresa Loveless, Courtney Carlson, Catalina Dentzel Helmy, Vincent Hu, Guohao Liang, Matt Demelo, Michelle Ficht, Arushi Singhai, Marcello Pajoh-Casco, Chang Liu

Our understanding of biology continues to deepen as our ability to determine the macromolecular content of cells and tissues increases in breadth and in resolution, which is now at the single-cell level for many experiments. Currently, techniques such as single-cell RNA sequencing only measure what is present at the time of destructive collection of the cells. DNA recording, in which each cell's history is recorded into its DNA, without destroying the cell, will eventually enable researchers to monitor the long-term activity of many signaling pathways in parallel, in many single cells, in physiologically-relevant contexts. As a contribution toward achieving that goal, we recently presented a DNA recorder called peCHYRON, which uses prime editor to record cellular events, in temporal order, at a genomic recording locus. peCHYRON requires neither synthetic recording loci nor double-strand DNA breaks, and is compatible with multiplexed recording of multiple types of events. Here, we present updates to peCHYRON that make it compatible with recording in multiple cell types and for a duration of approximately one month. As a demonstration, we show that peCHYRON can be used to reconstruct cell lineage and the order of chemical exposures. We put these current results in the context of the future work that will be necessary to increase the duration of recording and expand the breadth of signals that can be recorded.

AUTHOR INDEX

AIChE
120 Wall Street, FL 23
New York, NY 10005-4020

ISBN 978-1-7138-9302-8

CURRAN ASSOCIATES INC.
proceedings.com